A MATHEMATICAL APPROACH TO PROPORTIONAL REPRESENTATION:

Duncan Black on Lewis Carroll

A MATHEMATICAL APPROACH TO PROPORTIONAL REPRESENTATION:

Duncan Black on Lewis Carroll

edited by

Iain McLean
(Official Fellow in Politics, Nuffield College, Oxford)

Alistair McMillan
(Research Officer, Nuffield College, Oxford)

and

Burt L. Monroe
(Assistant Professor of Political Science, Indiana University)

Springer Science+Business Media, LLC

Library of Congress Cataloging-in-Publication Data

A C.P.I. Catalogue record for this book is available from
the Library of Congress.

ISBN 978-94-010-3735-8 ISBN 978-94-007-0824-2 (eBook)
DOI 10.1007/978-94-007-0824-2

Originally published by Kluwer Academic Publishers in 1996
Softcover reprint of the hardcover 1st edition 1996

Printed on acid-free paper.

Contents

Editors' Preface

This book has been thirty years in the making. As we explain in our Introduction, the main part of it is the posthumous work of the Scottish economist Duncan Black (1908-91). Black was an original and neglected thinker who had unusual difficulty in getting his work published. He was also a perfectionist. He told his friends that this book was ready to publish in the late 1960s, but he was never satisfied and continued to make additions and changes throughout the rest of his life. The small circle of people who knew the importance and originality of this work waited eagerly but in vain.

Black wanted Kluwer Academic Publishers, who had published the facsimile reprint of his classic *The Theory of Committees and Elections* (1958; Kluwer reprint 1987) to publish this book; but it was far from being in a fit state to publish at his death. Black was a widower with no immediate family, and at his death his books would have been sold and his papers dispersed had it not been for the intervention of Richard Alexander, an economist at the Royal Naval College, Dartmouth, Devon, and a neighbour of the agent charged with disposing of Black's real estate after his death. On finding that his house contained a substantial quantity of academic books and papers, Mr Alexander alerted Black's *almae matres* the University College of North Wales, Bangor (which declined them) and the University of Glasgow (which accepted them for its Archives). One of us (IM) visited Black's house with Zachary Rolnik of Kluwer. Amidst four roomfuls of miscellaneous papers, one of the first things we found was a chapter plan for this book. This helped us to decide that it was feasible to carry out Black's wishes. (Later, we discovered many other chapter plans, all different). The papers were hurriedly removed from Black's house and taken to Glasgow. Work on cataloguing them, with the help of a grant from the Economic and Social Research Council, began in 1993 and was completed in 1994. A grant from the Newlands fund of the University of Glasgow enabled BLM to join AM in Glasgow during the cataloguing. We were also assisted by Gordon Tullock (University of Arizona) and Gordon Brady (Sweet Briar College), who both visited Glasgow during the cataloguing process. The University of Warwick and Nuffield College, Oxford, successively administered the grants. Michael Moss and his staff at the University Archives in Glasgow were extremely generous and helpful hosts to the project.

Duncan Black was elusive and self-effacing. We are very grateful to Nick Baigent, Alec Cairncross, Ronald Coase, and Donald Stokes for their help in enabling us to learn more about him as a person. For expert advice on Lewis Carroll we wish to thank Morton N. Cohen. The Bodleian Library and the library of Christ Church, Oxford, contain some of our primary sources and we are grateful for their help.

Parts of the Introduction have appeared in the *Journal of Theoretical Politics*, 7(2), 1995. We are grateful to the publishers for permission to reproduce that material here.

To all those who have helped rescue Duncan Black's papers, we give our heartfelt thanks.

Oxford
Bloomington, Indiana
April 1995

Introduction

1. What is the Theory of Voting?

This book is about a well-known writer, Lewis Carroll, and about a little-known subject, the theory of voting. It has been edited from the manuscripts of a writer who is still not as well known as he should be, Duncan Black. There is no need to explain who Lewis Carroll was, but every need to explain who Duncan Black was, and what the theory of voting is, as it is a generally misunderstood subject.

It is best to begin by explaining what the theory of voting is not. It is not about the *content* of voting, nor yet about whether (and if so, how) a rational voter would ever vote. It is purely about voting *procedures.* A voting procedure has two main components. First, it prescribes how each voter should cast her vote: perhaps for just one candidate and perhaps for more than one; perhaps by ranking the candidates from best to worst; perhaps by dividing them into the two categories 'acceptable' and 'unacceptable'; perhaps for individual candidates, perhaps for a party list. Secondly, it prescribes how the individual votes are added up ('aggregated') to determine the group's choice, or the group's ordering. This may sound both routine and as dry as dust. It turns out that it is far from routine, and as a consequence that it is not at all dry.

There are a number of introductions to the theory of voting. Barry and Hardin (1982, Part II), Mueller (1989) and Saari (1994) are introductions to the concepts. McLean and Urken (1995) is an introduction to the history, with texts. Duncan Black's own *The Theory of Committees and Elections* (Black 1958) is an excellent introduction to both topics, up to the date when he wrote. Here, we will present only enough of the theory as is needed to provide a context for this book.

The theory of voting has two main subdivisions: *the theory of majority rule*, and *the theory of proportional representation.* 'Majority rule' has a clear meaning when only two candidates, or issues, are in contention. If one option wins more than half of the votes cast, it is unambiguously the majority's choice. Simple majority rule satisfies some classically desirable properties of fairness; furthermore, it is the only voting procedure that does so (May, 1952). But as soon as there are more than two candidates, simple majority voting may perform very badly. For instance, a candidate may be the plurality winner (that is, get the largest number of votes of any of the candidate, but fewer than half of the votes cast), while being the absolute majority loser (that is, ranked last of all by more than half of the voters). An obvious way to avoid this problem is to conduct exhaustive pairwise voting. Each candidate is compared with each other. If the voters are asked to rank the candidates, this may be done at the second stage in the procedure, the aggregation stage. There is no need for the voters actually to vote on each pair. Exhaustive pairwise voting was first proposed by Ramon Lull in 1283. If it yields a clear winner, that is a candidate who has won a majority against every other, taken one at a time, that candidate has an obvious claim to be considered the best. Such a candidate is called the *Condorcet winner* after the Marquis de Condorcet (1785), the greatest figure in voting theory before this century.

Unfortunately, the Condorcet winner may not exist. Whenever there are at least three candidates and at least three voters, there is a possibility that *A* wins a majority against *B*, who likewise beats *C*, who likewise beats *A*. Here is Lewis Carroll's statement of the problem (from Dodgson 1876, in Black 1958, p. 227):

> Let us suppose that there are 11 electors, and 4 candidates, *a*, *b*, *c*, *d*; and that each elector has arranged in a column the names of the candidates in the order of his preference; and that the 11 columns stand thus:

a	*a*	*a*	*a*	*b*	*b*	*b*	*c*	*c*	*c*	*d*
d	*d*	*b*	*b*	*c*	*c*	*d*	*b*	*b*	*b*	*c*
c	*c*	*d*	*d*	*a*	*a*	*c*	*d*	*d*	*d*	*b*
b	*b*	*c*	*c*	*d*	*d*	*a*	*a*	*a*	*a*	*a*

> Here the majorities are cyclical, in the order *a d c b a*, each beating the one next following.

The puzzled reader should check Lewis Carroll's arithmetic, which is correct. Candidate *a* beats *d* by 6 votes to 5; *d* beats *c* by 6 to 5; *c* beats *b* by 6 to 5; and *b* beats *a* by 7 to 4. When such a cycle exists, 'majority rule' seems to have no meaning: whatever society ends up with, a majority of voters would have preferred something (somebody) else. This has deep implications for democratic theory.

An obvious way round this difficulty is the rank-order count. This was possibly proposed by Lull in 1283 and certainly by Nicholas of Cusa in 1433, but is usually named after its second (or third) inventor, J.-C. de Borda (1733--99). The *Borda count* works as follows. Each voter ranks the candidates from best to worst. Where there are *n* candidates, this is done either by giving *n* points to one's favourite and so on down to 1 to one's least-liked, or from *n* - 1 at the top to 0 at the bottom. There are rules for equal places that need not concern us. These scores are simply added up, and the candidate with the highest aggregate score is elected. This is beautifully simple and, unlike the Condorcet procedure, always gives a clear result. But it is subject to paradoxes and manipulation of its own. It may fail to select the Condorcet winner, even when one exists; it may lead to peculiar outcomes when the set of candidates is expanded or contracted; and it is patently manipulable by voters putting the most dangerous rival to their favourite at the bottom of their lists. 'My scheme is only intended for honest men', said Borda when this was pointed out.

Lewis Carroll and Duncan Black both made important contributions to this theory. Carroll independently discovered both the Condorcet and Borda rules, and the problems listed in the last two paragraphs. He was the first to call the situation where there is no Condorcet winner a 'cycle'. The name has stuck. As Black explains in Part I of this book, Carroll's discovery was independent of Condorcet and Borda. The scientific journal containing Borda's paper is in the library of Carroll's college, Christ Church. But the pages containing Borda's paper are uncut. Nobody has ever read them in this copy. And this is the copy that Carroll would surely have read if he had read Borda. When Black visited Oxford in the 1950s, the same was true of the Bodleian Library's copy of Condorcet (1785), although it is no longer true. A full evaluation of Lewis Carroll's work on the problem of majority rule is in Black (1958). Much of it is also contained in Part I of this book. Black's view that Carroll deserves 'a position in the theory of elections and committees only a little lower than that of Condorcet' is now generally accepted.

In 1829 Felix Mendelssohn conducted the first performance of J S Bach's *St Matthew Passion* since Bach's death. What is now regarded as the greatest choral work of all time had been, in effect, lost for eighty years until Mendelssohn rediscovered it. Duncan Black was to Condorcet and Carroll as Mendelssohn was to Bach. He made two signal contributions of his own to the theory of majority rule: the median voter theorem, and the first statement of the problem of multidimensionality. He proved the median voter theorem while firewatching in Warwick Castle in 1942, although the proof was not published until 1948, with associated results (Black 1948a--d; Black 1991). The median voter theorem holds if the structure of opinion is single-peaked. (The reader who wishes to know exactly what single-peakedness is and how it gets its name should see Black 1958). Single-peakedness is guaranteed if the possible outcomes can be arrayed along one generally recognised dimension, perhaps from left to right. In industrial societies, there is a general consensus that high spending on social welfare is a 'left-wing' position, and giving priority to low taxation over high welfare spending is a 'right-wing' position. People recognise that this scale exists, whatever their own position on it. A left-wing person likes the leftmost option the most, and likes each succeeding option less the more right-wing it is. A right-winger has mirror-image preferences. Everybody else has some ideal point which is at neither extreme, and likes each alternative less, the more extreme it is, on each side of her favourite point. When these conditions obtain, the policy, or candidate, favoured by the median voter is the Condorcet winner, who (which) will be chosen by any good majoritarian voting procedure. The median voter is the voter whose has exactly as many people to the left of her as to the right. For simplicity, we assume that there are an odd number of voters.

Black's median voter theorem is a very powerful result, although it is often attributed rather to Anthony Downs (1957) who popularised it. It is now at the heart of empirical political science. For instance, it is now standard to model voting in the US Congress along the following lines. In any one policy area (say, protection v. free trade), there are three voting chambers: the House of Representatives of 435 members; the Senate of 100 members; and the unique President. Each is popularly elected, but by different voting procedures. Hence we may rank Representatives and Senators from the most protectionist to the most free-trading, and examine the position of the 218th most protectionist Representative, the 50th or 51st most protectionist Senator, and the President. These three could stand for the entire voting body as, in this simplified model, they are the voters on whom all depends. This is not the place to go into details (see, e.g., Cox and McCubbins 1993), but merely to note the central role of the median voter theorem.

However, the median voter theorem does not always work, as Black himself was one of the first to find out. When there is a cycle, there is no majority winner, and no median voter. When Black first discovered an example of a cycle,

> it seemed to me that this must be due to a mistake in the arithmetic. On finding that the arithmetic was correct and the intransitivity persisted, my stomach revolted in something akin to physical sickness. Not only was the problem ... more complicated than I had supposed, it was of a different kind (Black 1991, p. 262).

Often the reason why the median voter theorem fails is that opinion is structured in two dimensions at once. Voters may have to take a position not only on free trade v. protection, but also, at the same time, on high welfare spending v. low taxation. Voters' positions on one issue may not be related to their positions on the other, and the median voter on one may not be the median on the other. Black was the first to explore this systematically (Black and Newing 1951). This line of work led to some notorious 'chaos theorems' (e.g., McKelvey 1979) which seemed to show that in a complex society *all* possible outcomes were in a global cycle. This shows the depth of Lewis Carroll's cavern.

The theory of majority rule deals with the case where a group of people must choose the single 'best' candidate, or course of action. The *theory of proportional representation* deals with choosing an assembly that is in some sense 'representative' of those who choose it. 'Representative' has several meanings, with incompatible theories of representation attached to them. In one sense, an elected official represents those who elected her in the same way as a lawyer represents a client, or more generally as an 'agent' represents a 'principal'. The agent can choose only one course of action, so she must know what the majority of her principals want. This sense of representation is therefore bound up with the theory of majority rule. A radically different use of the same word is in the claim that an assembly should be 'representative' of those who elected it. Both the French revolutionary Mirabeau and the American revolutionary John Adams expounded this view during their country's revolution. For Mirabeau or Adams, an assembly should be some sort of map, or picture in microcosm, of the electorate (McLean 1991). The two main conceptions of 'representative' are both widely understood, but they cannot both be applied at the same time to the same situation. One of them is about the decisions an assembly reaches. The other is about its composition.

The second conception of representation in itself offers two incompatible answers to the question: But what sort of map - political or physical? In Europe and Australia, the theory of proportional representation developed as an analysis of what was required to ensure that the assembly was politically representative. In the USA, the concern was to ensure that it was physically representative. An assembly is politically representative if the parties are represented by the same share of seats as they each had of votes in the election to the assembly. An assembly is physically representative if each major subdivision of the population is proportionally represented. In the explosive history of proportional representation in the USA, this has mostly been taken as a question about fair representation of ethnic minorities. One of the issues thrown up by this is the fair apportionment of seats to subdivisions of a country such as states; another is the fair drawing of district boundaries once the apportionment has been made. Lewis Carroll's work on PR was mostly about political proportionality, although it opened with a discussion of physical proportionality.

There was a surge of interest in PR in Britain, Europe, and Australia in the second half of the 19th century. This led to a corresponding surge of writing, some of it by people whose names have become attached to some of the key concepts, such as Thomas Hare (1859, 1873), H R Droop (1868, 1881), Victor d'Hondt (1885), and A Ste-Laguë (1910). Most of these books and pamphlets contain arithmetical examples, and some of them contain general arguments. But of the writers who have given their names to concepts in PR, only Ste-Laguë argued in a mathematical way.

Below these (fairly) well-known names lies another tier of writers whose arguments are better but harder. These include C G Andrae (see Lytton 1864, Andrae 1926), E J Nanson (1882; see McLean and Urken 1995), James Garth Marshall (1853), Walter Baily (1869, 1872), and C L Dodgson (Lewis Carroll: Dodgson 1873, 1874, 1876, 1883, 1884, 1885). These writers may fairly be called mathematicians of voting. Andrae, Nanson, and Dodgson were professional mathematicians. (For the others, see Part 2.3 below). Each of them tried to treat the subject axiomatically, or made some approach to doing so. This approach took them into deep waters. They are deeper than most of those who write about PR, whether for it or against it, realise. They are deeper than most politicians realise. The theory of voting has had a strange history, and this book is itself part of that strange history.

2. Duncan Black and the Study of Theoretical Politics

Duncan Black (1908--91) was a Scottish economist who had a somewhat lonely career. He devoted his professional life to what he called 'the pure science of

Politics'[1] (Black 1991 p. 260). His formal theory of the committee is now the basis of an astonishingly wide range of work in both theoretical and empirical politics. A search through the on-line *Social Science Citation Index* and the *Arts and Humanities Citation Index*, covering major journals from 1981 to 1993 inclusive, reveals no fewer than 335 citations to Black (1958; 2nd impression 1968; facsimile reprint 1987); 86 citations to one or other of the seven papers which led to the 1958 book (Black 1948a-d; 1949 a-c); and 17 citations to the extremely elusive Black and Newing (1951). Furthermore, of this total of 438 citations, 43 (9.8%) were in 1993, which represents one-thirteenth (7.7%) of the time-span of the on-line indices. In other words, citation of these old works seems to be actually increasing over the years. If ever there were citation classics in social science, Black's publications from 1948 to 1958 must be among them. Downing and Stafford (1981) in their cross-sectional survey of social choice citations during 1978--9, found that Black (1958) ranked fourth with 25 citations, behind Arrow (1951), Sen (1970), and Downs (1957). As the obituary of Black in the *Journal of Theoretical Politics* 3 (3), 1991, p. 276, concludes, 'Only recently has he been recognised as a founding father of theoretical politics'.

Black's *Theory of Committees and Elections* (1958) is thus justly viewed as a classic. However, here is a very stark contrast between this and Black's later work. From 1962 until his death, he tried to explicate and evaluate Lewis Carroll's work on the theory of voting. During his lifetime, he published only four short papers on Carroll (Black 1953, 1967, 1969, 1970). In contrast to the 438 citations of Black's main work prior to 1958, there has not been a single citation, in any journal in the *SSCI* or *AHCI* sets, to these four papers. Nor has there been a single citation, in any form that we could trace, to Lewis Carroll's original papers on voting, written in the 1870s and 1880s. Black's papers on Lewis Carroll were early versions of a book to which Black devoted the last three decades of his life. At his death, copious drafts and chapter plans existed, but no book. We have reconstructed the book from Black's papers, as we describe in detail later. But first, we must describe the papers.

3. The Duncan Black Archive at the University of Glasgow

Duncan Black died in January 1991. Sadly but typically, it was some months before news of his death filtered through to the community which he had helped to found; in particular, he did not live to see the semi-autobiographical paper Black (1991) in print. However, in May 1991 we learnt from William H Riker, who, unfortunately, has also since died, that a clause in Black's will made some provision for the (re)publication of his work. He had left a substantial academic archive. On the initiative of a neighbour of the estate agent charged with disposing of Black's

---

[1] Black always distinguished between an activity and its professional study by capitalising the latter (see, e.g., Black 1958 p. 1). We preserve this characteristic idiosyncrasy.

property, this archive was offered to the University of Glasgow. Glasgow was Black's undergraduate *alma mater* to which he later returned as a Senior Lecturer in Economics. The University accepted the offer, but as no money was available to move it, let alone to catalogue it, nothing happened immediately. Black had left no dependants. (He was a widower without children). His house in Paignton, Devon was on the market, but found no immediate buyer as it was in poor condition: the price of its magnificent oceanside setting was a front garden that looked as if it was about to subside into the sea. One of us (IM) visited the house twice, once with a representative of Kluwer Academic Publishers and once with a representative of Glasgow University. Black's academic papers filled four rooms in his house, but, although he had been a meticulous archivist himself, they had inevitably fallen into some confusion before we saw them. We found a unique copy of Condorcet's first publication *Mémoire sur le calcul intégral* (1767), in a broken cardboard box full of slippers and sticking-plaster. Shortly afterwards, the archive was hurriedly removed from Black's house when it was finally sold. Sustained work on cataloguing it began in May 1993 and is now complete. The catalogue (McMillan 1994) may be obtained from the Archivist, University of Glasgow, Glasgow G12 8QQ, UK. It is hoped that a copy of all or part of the archive will be made on behalf of the Public Choice Center at George Mason University and housed there.

The archive is of considerable intrinsic interest. The principal components are:

notebooks and diaries, some going back to Black's days as a Glasgow undergraduate; Black bought up large quantities of expired page-per-day desk diaries, and most of the more recent notes are in these;

loose academic notes on various subjects in social choice, especially Lewis Carroll. Some are on paper issued by the University College of North Wales for students to write examinations, and it seems likely that Black wrote up some of his notes while doing the tedious job of invigilating examinations that falls to all British academics;

drafts of unpublished papers;

collections of offprints. These have proved unexpectedly useful in showing where Black's interests moved, and what he was researching at a time when nobody else was;

academic correspondence;

reminiscences and memoirs.

This archive is drawn on in the rest of this introduction. The catalogue is a Paradox database. All its contents are preceded by the general class mark DC 304, which is not repeated in the citations below.

4. Black's Struggles for Recognition

Black was born in 1908 in Motherwell, in the industrial belt of central Scotland some 15 miles from Glasgow. His father was one of the thousands of migrants from the rural Highlands who settled in the area; his mother was a Lowlander. Black benefited from the rigorous and formal education which was available to (some) clever working-class children in Scotland at the time. From Dalziel High School, Motherwell, he proceeded to the local university, Glasgow. There he graduated in mathematics and physics, which he did not enjoy. As he was drawn to social science, he re-enrolled for a second MA (= English or American BA) in economics and politics, from which he graduated with first class honours in 1932. His first academic job was in the Dundee School of Economics. This had been founded by a local industrialist in conscious imitation of the London School of Economics. The school throve as an independent research centre for a while before its (re-)absorption into the University of St Andrews, which had a campus in Dundee. In the early 1930s the school was a centre of innovative research. The two junior lecturers were Black and R. H. Coase, who formed a close friendship which lasted the whole of Black's life. Black was much influenced by his discussions with Coase, and stated that he sought a theory of the committee to parallel Coase's (1937) theory of the firm. Coase (1981) acknowledged Black's influence on his work but stated that Black's reciprocal acknowledgement to him was 'over-generous'. But we may be sure that the interaction between these two strikingly original economists (Coase is the 1993 Nobel Laureate in Economics) was fruitful. (For further biographical detail see Coase 1981, which was compiled in part from Black Archive 7/1, and Coase 1993).

Economics in Glasgow in Black's time reflected the breadth of interests of one of the earlier professors of it, Adam Smith. The drawback of this was that more breadth meant less depth. Black's Glasgow classmate Sir Alec Cairncross found going from Glasgow to do graduate work at Cambridge 'was like moving from a Kindergarten to a University' (Cairncross 1993; for more on Glasgow economics at the time see Cairncross 1992). Black, Cairncross and Coase all concur that Black's training as a technical economist occurred on the job in Dundee, not during his economics course in Glasgow. The compensating advantage was that Black was exposed to ideas from across the social sciences which were to shape his own work. In his writings he constantly reiterated the influences on him of A K White, C A Campbell, and (less frequently) W R Scott. These taught politics, philosophy, and economics respectively. White taught Black the virtues of Socratic method and abstraction from current affairs. Specifically, he inspired Black in the search for a 'pure science of politics'. Black found the tentative beginnings of this pure science in work by Mary Parker Follett, which was later reproduced by White himself (Follett 1918, pp. 24--5; White 1945, p. 95). Another book studied by Black and Cairncross in their politics class with White was Catlin's (1927) *The Science and Method of Politics* (Cairncross 1993). Mitchell (1993) has recently rediscovered Catlin's work.

As Mitchell points out, it could have been a foundation work of public choice if Catlin had had the required intellectual equipment. But instead, Catlin 'packed more and bigger bags [than other political scientists at the time] but never even went to the station'. Mitchell concludes (1993, p.452):

> Perhaps, then, it has been a good thing that the founders of public choice never read Catlin.... A science that cannot forget its founders is in grave danger.

The truth is that Black did read Catlin, Follett, and White, and carried out the task that they wished for but could not achieve themselves.

Details about Black's subsequent search for his pure science of politics are becoming better known (see especially Tullock 1981; Black 1991). He spent most of his professional career in the University College of North Wales, Bangor, with short spells at Belfast and Glasgow. This career pattern, of movement between Scotland, Wales and Northern Ireland but never to England, was not uncommon among Scots academics of Black's generation. The structure of university education in Scotland and England was so different that Black possibly never considered applying to work in England. The evidence is that he did not enjoy working in the isolated Bangor campus. Even at a much larger university, there would have been few colleagues to share Black's specialist interest; at Bangor, there was only one, the mathematician R A Newing. The median voter theorem came to Black while he was away from Bangor, 'in a flash... in February 1942':

> A little before it had been brought to my attention that my main effort during the preceding years had produced no tangible result. I was 'firewatching' in case of air raids, late at night in the green drawing-room of Warwick Castle.... Acting apparently at random, I wrote down a single diagram and saw in a shock of recognition the property of the median optimum. (Black 1991, p. 260).

His discovery of cycles some months later brought on, as we have seen, 'something akin to physical sickness' (Black 1991 p. 262). In the late 1940s and early 1950s, Black wrote a number of papers on majority rule and on proportional representation - see, e.g., Black 1949a, b. The most noteworthy finally appeared as Black and Newing (1951). This, as we have seen, was the first paper in the modern literature on multidimensional spatial voting (for an evaluation see Grofman 1981 pp. 36--38). It had a chequered history. Black and Newing submitted it to *Econometrica* in November 1949. It was immediately acknowledged. However, in September 1950 the editor's secretary wrote to Black about a change in submission procedure, with no comment on what had been happening to the paper in the meantime. Not until May 24, 1951 did William B Simpson write on behalf of the journal, accepting the paper for resubmission on condition that the authors acknowledged Arrow's results. Black was very angry at what he saw as a denial of priority, withdrew the paper, and published it separately through a small specialist publisher in the form of a 60-page booklet with the intimidating title *Committee Decisions with Complementary Valuation* (Archive, 4/ECOT/10--15). Black and Newing's dust-jacket blurb ran:

> The decisions reached by a committee must be related in some intelligible way to the opinions of the members. The authors of this book suppose that the members' opinions depend on two distinct sets of circumstances, i.e., that their valuations are complementary. Various mathematical techniques are employed to show the relationship that exists between the members' opinions and the decision that the committee takes. Some unexpected results are obtained.
>
> Whatever the merits or demerits of the book, it can safely be said that there is no other which has attempted to deal with this subject.

As self-promotion, this was characteristically Duncan Black rather than Hollywood. Some twenty copies of the book were still in Black's house at his death forty years later.

This incident was typical of a number throughout Black's career. Through various combinations of bad luck, editorial incompetence, Black's lack of an invisible college of colleagues in the UK, his modesty and his caution, hardly any of his work was published promptly. *The Theory of Committees and Elections* was rejected by several publishers, including Oxford University Press, whose reader entirely failed to understand it (Black to W H Riker, 13.3.61; 4/RIKE/2). A few other examples, from the Archive, may stand for many. In December 1948, Joseph Goldsen of RAND wrote to Black expressing interest in his work and asking for offprints which would be scrutinised by 'a group of American mathematicians and political scientists'. Black apparently asked the British Consul at San Francisco what sort of organisation RAND was. On receiving a reply from the British Embassy in Washington that 'the activities of the Rand Corporation are highly classified.... We understand that the United States Air Force would much prefer that, if you decided to respond to Mr Goldsen's enquiry, it should be communicated to the Corporation through themselves', Black seems to have let the matter drop (4/RAND/1--3). He apparently had no wish to be a cold warrior. The two countries in which Black's work was recognised were Italy and the USA. Black was welcomed by the emergent Italian public choice community because of his repeated emphasis on the Italian public finance school as a precursor of public choice (see, e.g., Black 1991 p. 260). At the end of his life he was much honoured by his election as a fellow of the Accademia Nazionale dei Lincei (4/ACAD/1--5). However, the publication of his work in Italy was equally ill-fated. He submitted a paper on Borda Condorcet, and Laplace to the *Giornale degli Economisti* in March 1951. It was accepted, but never published, and Black seems to have received no replies to his letters asking what had happened (4/GIOR/4--7). In a letter to Sir Geoffrey Keynes, asking whether he might be related to Maynard Keynes through a putative common ancestor, Black described himself wistfully as 'an economist ..., not a very good one, I admit, and I wondered whether I might have some distant connection with your late distinguished brother' (4/KEYN/1; 16.12.74).

5. Black and Lewis Carroll

Perhaps it was these experiences that drew Black to those other lonely and misunderstood people in his field, the Marquis de Condorcet and Lewis Carroll. The original drafts of *The Pure Science of Politics* comprised only what became Part I of *The theory of Committees and Elections*. As Black states, (1958, p. xi), the historical section on Condorcet, Borda, Laplace, Galton, Carroll and Nanson was added in the 1950s. He first discovered Condorcet (1785) in August 1948 (Black to A. de Pietri-Tonelli, 19.11.50; 4/PIET/1). This letter implies that in 1950 he had not yet discovered Carroll, whom he would have mentioned if he had. However, in the following year he made notes on Carroll's activities on the Governing Body of Christ Church, Oxford (2/1/13/1). In 1953 he published an account of his discoveries (Black 1953). Along with Robin Farquharson (see Farquharson 1969, p. 17)[2], who had discovered Carroll's pamphlets on social choice at Princeton, Black was responsible for rescuing Carroll's lost work. He also established (1958, p. 193) that Carroll's work was original, owing nothing to Condorcet or Borda. In *The Theory of Committees and Elections* Black presents a full account of Carroll's work of the 1870s on social choice, giving the political and (Black's interpretation of) the psychological context (Black 1958, pp. 189--213). But he ignores Carroll's other mathematical work on voting, namely his *Principles of Parliamentary Representation* (Dodgson 1884) and his 'Lawn Tennis Tournaments' (Dodgson 1883, to be found in Woollcott 1939). Black regarded them as unimportant, and the latter as 'quite trivial' (Black 1958 pp. 191, 213).

Quite soon after 1958, Black changed his mind. By at least 1962, he had settled on the view, which he held for the rest of his life, that Carroll's *Principles* was a very important work in social choice. In three papers explaining it and putting it in context, he made such extravagant claims as that it was 'the only work in Politics worthy of being placed no more than a single notch below that of Thomas Hobbes' (Black 1970, p. 1). Black had come to see that Carroll's work on proportional representation (PR) relied implicitly on game-theoretical arguments of Nash equilibrium. As game theory had not been invented and Nash not born, it was not surprising that Carroll's pamphlet had not been understood. Black devoted much of the rest of his life to this book, which was designed to explicate it. In the three papers and his drafts for this book, Black also advanced his psychological theories, linking Carroll's bouts of mathematical originality with emotional crises regarding

2 Farquharson's work also suffered long delays between writing and publication. His 1969 book was completed in 1958. His life was lonelier and more tragic than Black's or Carroll's: he became a schizophrenic drifter, and died in a fire on waste ground.

his child-friends. Thus in Black's view, the social choice pamphlets of the 1870s were linked with Carroll's feelings about the loss of Alice Liddell ('the' Alice), and his pamphlet of 1884 with the loss of Edith Denman, another child-friend whom Carroll visited in September 1884 to give a talk in the Derbyshire parish of her husband, a clergyman. In December 1884, in between the first and second editions of Carroll's *Principles*, Edith Denman died of complications arising from childbirth (Goodacre 1984). Other Lewis Carroll scholars did not share Black's view of the relationship between Carroll's mathematics and his emotions: Roger Lancelyn Green wrote, '[F]rankly I think you are barking up a Tum Tum tree' (Green to Black 18.2.69; 4/GREE/2)

Fortunately, however, Black was more interested in the intellect of C L Dodgson than in the psyche of Lewis Carroll. We discuss Black on Carroll on voting systems in the next section of this introduction. But, first, a word about Black's annotations of *Alice.* Martin Gardner's justly celebrated annotations of the *Alice* books and of *The Hunting of the Snark* (Gardner 1960, 1962, 1975, 1990) will be familiar to most readers. Gardner is a mathematician with a long-standing interest in the theory of voting and in quirky logicians. As theorists of voting have often been quirky logicians, these sets intersect. (On Carroll's equally eccentric predecessor Ramon Lull see Gardner 1983; McLean and London 1990; McLean and Urken 1995). Gardner expertly disentangles the logical and mathematical issues behind the jokes and puzzles in Carroll's children's books. He also gives entertaining and informative details about the national and local context. However, he missed two points which Duncan Black makes in this book: frogs in coal, and the characters in the *Snark.* On both points, we feel that Black makes a highly plausible case.

One of the fruits of Black's research in the papers of the Senior Common Room of Christ Church is reported in Part I below. In an envelope of press cuttings covering the years 1855 to 1862, he found three letters to *The Times* about frogs in coal written in September 1862, while Carroll was expanding his original tale of Alice's adventures for publication. Although the letters have nothing to do with proportional representation, Black included them in his book draft. They are a marvellous urban myth, to correspond with today's 'baby in a microwave' and 'block of ice from an aircraft toilet' versions. As with modern urban myths, each writer testifies solemnly that he knows somebody of unimpeachable honesty who actually saw a live frog (or toad) jump out of a slab of coal (or marble) when it was cut. Further authenticity is given by the deadpan authority of the letter-writers' addresses. Who could argue with a man writing from Lilleshall Coal Depots, Great Western Railway, Paddington? As Black says, the correspondence between the frogs in these three letters, and the Frog Footman in the 'Pig and Pepper' chapter of *Alice in Wonderland* is remarkably close. We hope that the letters about frogs in coal will find their way into the next edition of *The Annotated Alice.*

More directly connected with the theme of this book is Black's conjecture about the Bellman and the Beaver in *The Hunting of the Snark.* Lewis Carroll frequently repeated that

> periodically I have received courteous letters from strangers, begging to know whether 'The Hunting of the Snark' is an allegory, or contains some hidden moral, or is a political satire: and for all such questions I have but one answer, 'I don't know' (written in 1887; quoted by Gardner 1962, p. 17, and by Black in Part I below).

However, at the end of his life Carroll conceded that there might be a subconscious allegory in the story: 'the [meaning] I like best (which I think is partly my own) is that it may be taken as an allegory for the pursuit of happiness' (written in 1897; quoted by Gardner (1962), p. 18). Numerous writers have taken this as their cue to identify the people Carroll was thinking about when he drew the characters in the *Snark.* Both Martin Gardner and the psychoanalyst Phyllis Greenacre thought that Carroll subconsciously drew himself in the tragic hero, the Baker, who forgets who he is and who softly and suddenly vanishes away (for the Snark *was* a Boojum, you see: Gardner 1962, pp. 55, 72). In Part I below Black argues for another interpretation: Lewis Carroll is not the Baker, but the Beaver.

> There was also a Beaver, that paced on the deck,
> Or would sat making lace in the bow:
> And had often (the Bellman said) saved them from wreck,
> Though none of the sailors knew how.

Black's conjecture about the Beaver hangs together with his conjecture about the Bellman, which rings true. As Black baldly says, the Bellman is H G Liddell, the Dean (i.e., president) of Christ Church, Alice's father. The persuasive subconscious connection between Liddell and bells is that while Carroll was writing *Snark,* Dean Liddell was pressing a new design for a belfry on the college. The controversy over the belfry gave rise to Carroll's first two pamphlets (Dodgson 1873, 1874) on the theory of voting, which represented the first axiomatic writing in English on the subject. Black suggests that by proposing voting procedures which helped the college out of its impasse over the belfry, Carroll had indeed saved them from wreck, though few of the sailors knew how. As for the Bellman, if Black's conjecture is right, the verses he quotes are a truly savage picture of Dean Liddell as an unctuous authoritarian ('What I tell you three times is true'), who had no idea where he was going ('they shortly found out/ That the Captain they trusted so well/ Had only one notion for crossing the ocean,/ And that was to tingle his bell'). As Black points out, such savagery is consistent with Carroll's other Oxford pamphlets of the same period.

6. Carroll's *Principles of Parliamentary Representation*

In Carroll's day, the widening of the franchise in the United Kingdom was highly topical. Conservatives might concede that it was hard to resist the principle, yet feared that they would lose out substantially. The franchise was reformed in 1832 and 1867. The 1867 Reform Act was noteworthy for a provision that some of the large cities should each form a three-member district, with voters having only two votes each. This was known as the 'limited vote', and it was introduced to the bill by a hostile Conservative amendment in the House of Lords which the Prime Minister, Benjamin Disraeli, unexpectedly accepted. The impact of the limited vote on Carroll's thinking is discussed below.

The British General Election of 1880 saw the closest approach since 1841 to a straight two-party contest in England, Scotland, and Wales (Cox 1987, especially Tables 9.4 and 9.5; McLean 1992). It was therefore the best illustration for forty years of the exaggerative effect of the relative majority (first-past-the-post) electoral system. For some years since its rediscovery by Kendall and Stuart (1950), this has been known as the cube law, because in the circumstances that obtained in 1880, the ratio of seats won by the two largest parties can be expected to be the cube of the ratio of their votes. (For a more general formula giving estimates to cover a wider range of circumstances see Taagepera and Shugart (1989), chaps. 14 and 16). The Marquess of Salisbury, leader of the Conservatives, noted this effect for the general election of 1880, saw that the Liberal lead in votes had produced an exaggerated lead in seats, and claimed that unless the 1884 extension of the franchise was accompanied by a redistribution of seats the result could be the destruction of the Conservatives in parliament even if their share of the vote was reduced only slightly or not at all by the franchise extension. He showed that if the electorate of a seventeen-seat legislature with single-member districts was split between imaginary parties which he named 'Catholics' and 'Liberals' in the proportion eight to nine, there were two circumstances in which the 'Liberals' would win all seventeen seats: where the population was exactly evenly mixed, and where it was completely segregated (say into a 'Liberal' city surrounded by 'Catholic' countryside), but constituencies were drawn in such a way (in this case, radially from the city centre) that each constituency contained the same ratio of 'Catholics' to 'Liberals' as the population (Salisbury 1884).

Ireland was even more threatening. Since 1874 and especially since 1880, seats in Catholic Ireland had been falling to militant supporters of Home Rule, who used every procedural means open to them to disrupt Parliament. The franchise reform of 1884 proposed to extend the franchise in Ireland, as in the rest of the country, to rural householders. Would this not mean a great boost to Parnell, the Home Rule leader, with consequent threats to public order and the unity of the UK? The uncomfortable truth was best pointed out by Sir John Lubbock, one of the saner advocates of PR, in the autumn of 1884:

> At the general election of 1880, 86 seats were contested. Of these the Home-rulers secured 52, the Liberal[s] and Conservatives together only 34. Yet the Home-rule electors were only 48,000, while the Liberals and Conservatives together were no less than 105,000. . . . we are told . . . that under the new Redistribution Act the Home-Rulers will secure 90 seats out of 100, leaving only a dozen to the Liberals and Conservatives together. . . .

> out of Ulster it is probable that scarcely a single Liberal or Conservative member will be returned. The result of this system, then, will be that Ireland will be entirely misrepresented, and that we shall have gratuitously created serious and unnecessary difficulties for ourselves. To adopt, indeed, a system of representation by which we shall exclude from the representation of Ireland one-third of the electors, and give almost the whole power to two-thirds, would, under any circumstances, be unjust; but to do so when the one-third comprise those who are moderate and loyal, while the two-thirds are led by men not only opposed to the Union, but in many cases animated by a bitter and extraordinary hatred of this country, seems to be an act of political madness. (Lubbock 1885, pp. 20--21).

Lubbock went on to draw an analogy from the US Presidential election of 1860, in which Abraham Lincoln won an absolute majority of the Electoral College on less than 40 per cent of the vote. Lubbock's prediction was exactly correct. The Home Rulers won eighty-five seats in the ensuing General Election and continued to do so at every election until 1910. Anglo-Irish war was about to break out in 1914 but was delayed for the First World War to take place. The Irish war lasted from 1919 to 1922, and some would say it has not yet ended (although 1994 has seen a truce, and possibly peace). Thus electoral systems have important consequences.

This then is the anxious context in which Carroll wrote his *Principles*. Carroll was a political Conservative as well as a temperamental conservative. He met Salisbury and his family in 1870, for once (uncharacteristically) using his fame as the author of *Alice* to obtain an introduction to Salisbury's wife and daughters (Cohen 1979, p. 211). Despite the gulfs of class and temperament, Carroll was welcomed by the Salisbury family and spent the New Year at their great house, Hatfield, several times in the 1870s and 1880s. Carroll seems to have thought about PR for the first time in 1882, in connection with college politics (see his diary entry for 17 May 1882 in Green 1953, pp. 405--6), but it was the reform crisis of 1884 that brought him into print. He wrote several letters on it to the *St James's Gazette*, his ideas evolving continuously. In June he hit on the most distinctive feature of his scheme, 'the giving to each candidate the power of transferring to any other candidate the votes given for him' (Diary for 3 June 1884 in Green 1953, p. 426). In July he sent it to Salisbury saying 'How I wish the enclosed could have appeared as *your* scheme. . . . That *some* such scheme is needed, and much more needed than *any* scheme for mere redistribution of electoral districts, I feel sure.' Salisbury replied immediately, acknowledging the need for electoral reform but stressing the difficulty of getting a hearing for 'anything . . . absolutely new. . . however Conservative.' Carroll replied the next day. After congratulating Salisbury for a speech in the House of Lords in which he had insisted that the Conservatives would not accept franchise reform unless it was linked with redistribution, Carroll went on '*please* don't call my scheme for Proportionate Representation a 'Conservative' one.... *all* I aim at is to secure that, *whatever* be the proportions of opinion among the Electors, the *same* shall exist among the Members' (the three letters are partly quoted and partly paraphrased in Cohen 1979, pp. 544--5). The House of Lords did indeed return the franchise bill to the Commons with an added clause insisting that it must be 'accompanied by provisions for so apportioning the right to return members of parliament as to insure a true and fair representation of the people' (quoted by Hart 1992, p. 107). This amendment was moved by Earl Cairns, the same peer who had inserted the limited vote provision into the 1867 Reform Act. We do not know how far the Lords' motives - those of the majority Conservative peers, in particular - reflected self-interest and how far a desire for fair representation. It may be pointless to try to separate the two, as fair representation was expected to mean, in Britain, protecting the Conservatives from being wiped out in terms of seats in a General Election where they came narrowly second in terms of votes. Salisbury may have wished to preserve the limited vote (Hart 1992, p. 111), but in the end went down a different road.

Like Sir John Lubbock, Carroll saw that Salisbury had failed to accept the implications of his own argument. No redistribution that retained single-member districts with the plurality voting rule could be guaranteed to save the Conservatives in Britain or either of the British parties in Ireland. As Salisbury (1884) himself pointed out and as Carroll repeated in his *Principles of Parliamentary Representation*, single-member districts combined with an even distribution of supporters of two parties around the country could lead to the larger of the two wiping out the smaller in terms of seats. It seems that Salisbury could not shift his perspective from majoritarian to proportional in order to see the true implications of his own argument. At this point, the concepts of political and physical proportionality interact with one another. Conservative supporters were fairly evenly spread around Great Britain; Irish Nationalists were heavily concentrated in Ireland. Salisbury's mind turned to a scheme which achieved physical proportionality in Britain (although not in Ireland). He masterminded the scheme of single-member districts 'according to the occupation of the people' that was embodied in the Redistribution of Seats Act 1885. This scheme, the outcome of bipartisan negotiations in the autumn of 1884, has frequently been hailed as Salisbury's stroke of genius (e.g., by Cornford 1963). It constructed suburban seats where the new concentration of Conservative voters was to be found in what contemporaries called the 'villa vote'. However, we cannot say whether it would have saved the Conservatives if the debacle Salisbury feared had come about. For in 1886 the Liberals were torn apart over Irish Home Rule, leading to a twenty-year Conservative hegemony not foreseen by Salisbury or anybody else in 1884. Thus no Conservative, in the end, had to take Carroll's arguments seriously out of self-interest.

It is unfortunate that points that Carroll took for granted and passed over quickly were exactly the ones that mainstream politicians could not accept, even when it was in their own interest. Carroll takes for granted both that guaranteeing the survival of minorities in parliament requires multi-member districts and some form of minority representation (which the politicians should have accepted but did not) and that the number of electors per MP should be equal (which almost no parliamentarian in the 1884 debates did). This was probably enough on its own to blind contemporaries to the more striking features of *The Principles* and may have misled Black into his initial dismissal of them

The Principles is the earliest known work to discuss both the assignment of seats to each of a number of multi-member districts (the apportionment problem) and the assignment of seats within each district to the parties (the PR problem). Not until Balinski and Young (1982) was the formal similarity between the two problems fully understood. Carroll largely understood, although he makes one mistaken statement in his section on apportionment, saying that electoral equality requires

> That each Elector, who is represented at all, should be represented by the same fraction of a Member. Or (which is the same thing) that each Member should represent the same number of Electors (Dodgson 1884, p. 3).

But minimising variance in fractions of a seat per elector is not the same as minimising variance in electors per seat. To minimise the first, an arithmetic-mean divisor rule is needed; to minimise the second, a harmonic-mean divisor rule is needed (for explanations, see Balinski and Young 1982; McLean and Mortimore 1992). Apportionment, like social choice, is a subject that has been periodically discovered, lost, and rediscovered. It seems likely from Black's collection of offprints that he was on the trail of this lost literature. For during the 1960s he had collected papers by the Harvard mathematician E V Huntington, who in the 1920s had produced what was then thought to be the optimal scheme for apportionment of seats in the House of Representatives to the states. Huntington was the founder of the modern study of apportionment, and also a little-recognised precursor of Black's great rival Kenneth Arrow. Huntington proposed (what is now known as Arrow's axiom of) Independence of Irrelevant Alternatives as a criterion for aggregation rules (Huntington 1938; McLean 1995). Black was also reading Sawyer and MacRae (1962) and possessed a copy of other papers on voting procedures in Illinois, including Moore (1920) (c.f. also Black Archive 5/2/BLAI/1; 5/2/HYNE/1). These papers discuss the cumulative voting scheme introduced in Illinois by its 1870 Constitutional Convention. Black, in his annotations, notes the similarity between the Illinois scheme and Carroll's proposals. These schemes have not been generally studied by social choice writers.[3]

For the assignment of seats to parties, Carroll recommends the standard ('Droop') quota $Q_d = V/(S + 1)$, rounded up to the next integer, where V = total votes cast and S = number of seats to fill, though he does not cite H R Droop and there is no reason to suppose he read him. He parts company with the Proportional

3 Goldburg (1994) has rediscovered Sawyer and MacRae (1962) and notes how that paper is a pioneer of game-theoretic analysis of a voting procedure. Her paper follows similar lines to Black's exposition of Lewis Carroll in the main part of this book. She cites a few other writers who have used Sawyer and MacRae's data, but not either Carroll or Black.

Representation Society, whose literature he does cite, because he shows that its rules (still the standard rules for Single Transferable Vote) could lead to the defeat of a candidate who had obtained a Droop quota. This cannot happen at the first stages of an STV election, but Carroll shows that it can happen at subsequent stages. Indeed, though Carroll is not at all in a Condorcetian mood in *The Principles*, he còuld have pointed out that in his example STV elects the Condorcet loser among the last three candidates. He briskly concludes that he has 'sufficiently proved the fallacy of its method for disposing of surplus votes.... Clearly *somebody* must have authority to dispose of them: it cannot be the Elector (as we have proved); it will never do to refer it to a Committee. There remains *the Candidate himself, for whom the votes have been given*.' Some may find this reasoning too much like Lewis Carroll's rather than Charles L Dodgson's, but if it seems bizarre, note that it flows from an assumption that Carroll *shared* with the STV school and that precluded Carroll from seeing the problem in a way in which either the Borda winner or the Condorcet winner would have been relevant. From a social choice perspective, the root problem of STV and all other elimination systems is that they use information about voters' preferences other than their first in an arbitrary way. The $n + 1^{\text{th}}$ preference of a voter whose n^{th} preference has been eliminated is counted. The $n + 1^{\text{th}}$ preference of a voter whose n^{th} preference has been elected with a surplus is counted with reduced weight, and the $n + 1^{\text{th}}$ preference of a voter whose n^{th} preference is elected with nothing to spare is not counted. Thus preference orderings are not treated equally. When writing about majority rule, Carroll had shown himself well aware about this; when writing about PR, he did not consider it.

Having, as he sees it, disposed of the conventional case for PR, Carroll goes on to his constructive argument, which is both compressed and elliptical. As Black was the first to see, it is essentially an argument to establish the Nash equilibrium strategy for two parties. Carroll considers the class of methods in which voters may each cast v unranked ballots in an m-member district ($v \leq m$). This was topical because of the 'limited vote' which operated in Manchester, Birmingham, Liverpool etc. with $v = 2$ and $m = 3$, and in the City of London with $v = 3$ and $m = 4$. In Birmingham, the Liberals had manipulated the limited vote by dividing the city into three zones, and asking their supporters in each zone to vote for a different pair of the Liberal candidates. They thus won all three seats in each of the general elections in the period. In 1874 the Conservatives did not run at all. In 1868 the Liberals controlled 73.1% of the votes cast, and in 1880 they controlled 67.2%.[4]

Were the Birmingham parties rational strategists? This may be answered directly from Carroll's pamphlet, once his reasoning is understood. Carroll was aware (but unfortunately failed to make explicit) that, given perfect information and

4 Source: Craig 1977. In 1868 and 1880 the Liberals put up three candidates and the Conservatives two. The number of Liberal voters was calculated by dividing the sum of the votes for the three candidates by two; the number of Conservative voters was taken as the number of votes for the more successful of the two Conservative candidates.

common knowledge about party strengths, the maximin (and Nash equilibrium) strategy for each party is to put up exactly as many candidates as it can fill seats if it instructs its supporters to divide their ballots among its candidates as evenly as possible. Carroll aims to find the voting procedure which leaves the fewest voters 'unrepresented', that is, whose votes do not contribute to the outcome. Carroll took this idea from the PR literature of his day which, then as now, sought to minimise the number of votes which a procedure 'wasted'. His conclusion is that, for the class of voting procedures he considers, the fairest - in the sense that, *ex ante*, it leaves the fewest voters 'unrepresented' - is that where $v = 1$ for any m, and that the fairness of the system increases with m. The limited vote had been introduced by the anti-democrat Lord Cairns, but Carroll had found a democratic justification for it. In other words, the fairest of this class of systems is that which divides the country into multi-member constituencies in which each voter has only one vote. This is the system used in Japan for national elections from the end of the Second World War until 1993, and generally labelled Single Non-Transferable Vote. SNTV is also used in Taiwan and formerly in Korea (Cox and Niou 1994); a limited vote system is used for the Spanish Senate (Lijphart, Lopez Pintor and Sone 1986). Note that Carroll's concept of 'fairness' is an important one, which is not liable to the criticism levelled by, e.g., Dummett (1984, pp. 214--5, 278--80) and Barry (1986, pp. 88--128) against the proponents of STV. As used in controversy about the merits of proportional representation, the concept of a 'wasted vote' is undefined because, as Dummett and Barry point out, *ex post* it cannot be determined which votes were wasted. Carroll's stochastic ex ante definition is important and should be dusted off.

If we look up Carroll's table (see p. 160 below), we find that for $v = 2$ and $m = 3$, a party with at least 61% of the vote can guarantee to win all three seats. Thus in Birmingham the Liberal strategy was optimal. Conservative strategy was optimal in 1874, and suboptimal in 1868 and 1880, at any rate in the narrow perspective of Carroll's game. If it was common knowledge that the Liberals controlled over 60% of the probable voters, there was no point in the Conservatives' running candidates, unless for the purpose of forcing an election and putting the Liberals to some expense.

Alice in Wonderland and *Through the Looking-glass* are based on games, the latter being explicitly a game of chess. Black was the first to see that Carroll's writings about voting were also about games. Carroll's life was devoted to being serious about games and gamelike about serious things. Occasionally Carroll actually refers to a 'game' (see., e.g., Black 1958, pp 232--3). But his more remarkable achievement is to have written about voting in game-theoretic terms, before game theory had been invented. One important paper, 'Lawn Tennis Tournaments', is about games in both senses. It shows how the conventional knock-out competition[5] is a dreadful way to rank players, as all players beyond the first-

---

[5] Also known in sporting parlance as 'single elimination tournaments', such competitions are quite familiar. Contestants are paired and eliminated after their first

place winner may achieve their places more through a fortunate draw than through skill or good play. Carroll's discussion anticipated concepts that appeared much later in the literature on social welfare functions and choice from tournaments (Miller 1980; Moulin 1988; on 'Lawn Tennis Tournaments' itself see Knuth 1973, pp. 209--11).

Carroll's result on SNTV has been independently rediscovered in a series of papers by Gary Cox and collaborators (Cox 1991, 1993; Cox and Niou 1994; Cox and Rosenbluth 1993, 1994). Under- and over-nomination in the limited vote in Japan and Spain are discussed by Lijphart, Lopez Pintor and Sone (1986). Cox's approach differs from Carroll's in that he postulates rational voters rather than rational parties (although in Cox and Niou 1994 the focus is on rational parties, in exactly the same way as in our and Carroll's discussions of Birmingham). But the outcomes of a procedure in which voters seek to avoid wasting their votes and of one in which parties seek to minimise waste of 'their' votes are essentially the same.

7. The Making of This Book

As stated above, Duncan Black originally believed that Lewis Carroll's work on PR was much less important than his work on majority rule. Within a short time of publishing *The Theory of Committees and Elections* in 1958, however, Black had changed his mind. He started to collect material for papers and a book on Lewis Carroll's theory of proportional representation. Three of the papers were published, attracting no attention at all (Black 1967, 1969, 1970). However, the book was never finished. Indeed, it became a kind of Penelope's web, occupying the last 30 years of Black's life. He described it as almost finished in the late 1960s. But he conducted a 20-year argument with himself on what to leave out and what to put in. He wrote down plans and sections as they occurred to him, usually dating them as he wrote in the out-of-date page-per-day desk diaries that he used as notebooks. Several times we have transcribed documents with care, only to come across a dispiriting message at the end such as 'This is all rubbish'.

Nevertheless, the outline of the intended book was always clear. In the first search of Black's house after his death, one of us (IMcL) and Black's editor at Kluwer Academic Publishers came across a chapter plan for the book. Discovering the chapters themselves took much longer, but from that day we were sure that there was a book in the four roomfuls of material that was removed from Black's house - if only we could find the component parts. Later, the problem turned from shortage to glut. There were many, many different versions of most of the parts, and we were faced with the tricky task of selecting the best. A common device in this situation is to look for the latest in each case, as representing the author's final thoughts. This was always too neat for our problem, and we suspect it is too neat in many other

loss. First-round winners are paired and again losers are eliminated. This continues until a single contestant remains as the unbeaten winner.

cases as well. An author does not usually know exactly when he is going to die. Which of us, in returning to revise our work, steadily improves it at every revision until the arbitrary day we die? Certainly not Duncan Black, on his own admission. He went through spells of physical illness and mental depression when he felt certain that he was not working at his best - and we agree with him. On the other hand, consistently choosing the *earliest* version of each part does not work either. During his many years of work on Lewis Carroll, Black made clear discoveries and improvements.

Therefore, no mechanical rule for selecting versions was available for us. We have had to use our editorial judgment to select what seems to us the best - most coherent, most concise, most comprehensive, as circumstances seemed to demand - version of each section. Anybody who wishes to re-examine our editorial judgments has the same primary material as we had. All the versions we have found of every chapter - from substantial typescripts to short notes on torn pieces of paper - are in the Duncan Black Archive, and catalogued in McMillan (1994). We have also tried to be as unobtrusive as possible. Everything in Parts I--III below is by Duncan Black, except where we have marked essential additions by square brackets [ ]. We do not always agree with Black's interpretations. Some of our disagreements are in this Introduction. In other cases, we have left it up to the reader to evaluate Black's interpretation. A statement that we disagreed with it would merely be setting the opinions of three people who know something about the subject of the book against those of the one person who knew more about it than anybody else.

All of Black's chapter plans made it clear that the book was to be in three parts written by himself, followed by a reprint of Carroll's *Principles of Parliamentary Representation* and its main sources, as Black saw them. Part I is biographical, introducing Lewis Carroll and giving relevant details of his life. It includes a few things, such as the cuttings about frogs in coal, which are nothing to do with proportional representation. But to have cut them out would have been puritanical. Black intended this part to be made up from lectures he had given at various American campuses, including the University of Virginia and Tulane University, and from the biographical material he had already put into *The Theory of Committees and Elections*. We had to include some of the latter for the sake of the flow of Black's argument, and to make this book free-standing, but we have tried to minimise material duplicated from the earlier book. Part I also contains Black's exploration of Lewis Carroll's logical mind through the records he kept as Curator of the Senior Common Room at Christ Church. Because of its origin as lectures, some of Black's presentation in this part is quite informal, but we have not attempted to change this.

The core of Part II is Black's already published work on Lewis Carroll (Black 1967, 1969, 1970). We have edited it to bring in Black's later thoughts, put the three papers together in what seemed to be the most logical order, and eliminated repetition as far as possible.

Part III presented our greatest challenge. It comprises the more detailed arguments about Carroll's reasoning, and his relationships with earlier writers on PR, that Black intended to use to supplement Part II. There are many plans in the Black papers for the intended arrangement of the sections within this part. They are not all the same, and we have not been able to follow any of them slavishly, partly because of missing material. Again, we have done our best to produce a coherent and flowing argument, taking the best version of each piece that we could find.

Part IV will, we hope, be useful independently of the rest of the book. The texts we reproduce are extremely scarce. None of them had seen the light of day this century until McLean and Urken (1995), which reprints *PPR* but not the others. They may be read along with Jenifer Hart's *Proportional Representation: critics of the British electoral systems 1820--1945* (Hart 1992).

We hope that we have shown that Duncan Black, who was educated in the department where Adam Smith once taught, was as broad in his interests as his great predecessor. (There are no surviving Smith materials in the University of Glasgow Archives, and the staff there have been exceptionally helpful in ensuring that Black does not meet the same fate). Some of the contents of the archive are drawn on in this book; others, we hope, will appear in a new edition of *The Theory of Committees and Elections* and perhaps in another volume of Black's unpublished papers. But there is much which will be in none of the books but has been preserved in the archive. We hope that other users will find it as fascinating as we have.

Black obviously had deep empathy with Lewis Carroll. His own attitude to sex was as puritanical as Carroll's (and he worried greatly whether Carroll ever did anything improper with his child friends despite the lack of any evidence that he did). He occasionally criticises Carroll - for the savagery of his attacks on Dean Liddell in Part I below, and for his failure to acknowledge James Garth Marshall and Walter Baily in Parts II and III. But there is more extravagant praise than criticism in the pages that follow. This book is the second of Black's sustained efforts to restore the reputation of Lewis Carroll as a serious logician. It complements the work of others who currently aim to rescue Carroll's reputation in other branches of his mathematical work (especially Seneta 1984, 1993; Abeles 1993; Knuth 1973, pp. 209--11). But it is more than that. It is the tribute of one misunderstood and lonely scholar to another.

Many Carroll scholars have commented on the poignant tone of Carroll's comments on the grown-up Alice in *Through the Looking-Glass* (published when she was nineteen). Carroll now saw himself as 'half a life asunder' from Alice in the dedicatory poem[6] and, parodying himself as the White Knight, took his farewell from her in chapter VIII:

---

[6] Did A E Housman have this line of Carroll's in mind when he wrote of his own wrenching separation from Moses Jackson in 1898?:

He would not stay for me; and who can wonder?
He would not stay for me to stand and gaze.
I shook his hand and tore my heart in sunder
And went with half my life about my ways.

> So she went on talking to herself, as she watched the horse walking leisurely along the road, and the Knight tumbled off first on one side and then on the other. After the fourth or fifth tumble he reached the turn, and then she waved her handkerchief to him, and waited till he was out of sight. 'I hope it encouraged him,' she said, as she turned to run down the hill....

Duncan Black felt for Lewis Carroll what Alice felt for the White Knight.

But the fate of Black's work to date raises broader questions. How could so original a thinker be so scandalously neglected in his lifetime? Admittedly, Black's work has been given something like its due in the U.S.A. and in Italy, but it remains almost unknown in his native land. Academe is not always kind to innovators. Just as James Watson (who was supposed to be researching polio) and Francis Crick (who was supposed to be a physicist) were warned off working on DNA in the early 1950s, so Black reports that in 1942 'it had been brought to my attention that my main effort during the preceding years had produced no tangible result' (see above). The theory of voting is marked by repeated inventions, losses, and rediscoveries. For instance, strategic voting was discussed by Pliny the Younger in AD 105. A matrix method for pairwise comparison among more than two options was proposed by Ramon Lull in 1299 (McLean and London 1990). The Borda count was invented by Nicholas of Cusa, ca. 1435. The fundamentals of the theory of voting were laid by Borda, Condorcet, and a few others in late 18th-century France, only to be lost completely until Black rediscovered them. The same is true for Carroll's writings of the 1870s and 1880s. (McLean and Urken 1995 collects all these texts together). In modern times, economists have neglected the theory of voting as it is only barely part of economics, while political scientists have neglected it mostly out of sheer ignorance. At least Black's own bad luck seems to have made him sympathise with his equally unlucky predecessors Condorcet and Carroll. For that we should be grateful, as it led to Black's fine work in the history of social choice.

References for the Introduction

Abeles, F. (ed.) (1994) *The Mathematical Pamphlets of Charles L. Dodgson* Charlottesville, VA: University of Virginia Press.

Andrae, P. (1926) *Andrae and his Invention, The PR Method* (Translated by V. Meisling) Copenhagen: P. Andrae.

Arrow, K. J. (1951) *Social Choice and Individual Values* New York: Wiley.

Housman, *Additional Poems* VII; see Page (1983), pp. 86, 184--5.

Baily, W. (1869) *A Scheme for Proportional Representation* London: Ridgway.

Baily, W. (1872) *Proportional Representation in Large Constituencies* London: Ridgway.

Balinski, M. L. and Young, H. P. (1982) *Fair Representation* New Haven: Yale University Press.

Barry, B. (1986) 'Wasted votes and other mares' nests: a view of electoral reform'. Pasadena, CA: Caltech Social Science Working Paper 612.

Barry, B. and Hardin, R. (eds.) (1982) *Rational Man and Irrational Society? an introduction and sourcebook* London: Sage.

Black, D. (1948a) 'On the rationale of group decision-making', *Journal of Political Economy* 56: 23--34.

Black, D. (1948b) 'Un approcio alla teoria delle decisioni di comitato', *Giornale degli Economisti* May--June: 1--23.

Black, D. (1948c) 'The decisions of a committee using a special majority', *Econometrica* 16: 245--61.

Black, D. (1948d) 'The elasticity of committee decisions with an altering size of majority', *Econometrica* 16: 262--70.

Black, D. (1949a) 'The theory of elections in single-member constituencies', *Canadian Journal of Economics and Political Science* 15: 158--75.

Black, D. (1949b) 'Some theoretical schemes of proportional representation', *Canadian Journal of Economics and Political Science* 15: 334--43.

Black, D. (1949c) 'The elasticity of committee decisions with alterations in the members' preference schedules', *South African Journal of Economics* 17: 88--102.

Black, D. (1953) 'Discovery of Lewis Carroll documents', *Notes and Queries* 198 (2): 77--9.

Black, D. (1958) *The Theory of Committees and Elections* Cambridge: Cambridge University Press.

Black, D. (1967) 'The central argument in Lewis Carroll's *The Principles of parliamentary representation*', *Papers in Non-market Decision-making* 3: 1--17.

Black, D. (1969) 'Lewis Carroll and the theory of games', *American Economic Review, Proceedings* May: 206--10.

Black, D. (1970) 'Lewis Carroll and the Cambridge mathematical school of P.R.; Arthur Cohen and Edith Denman', *Public Choice* 8: 1--28.

Black, D. (1991) 'Arrow's work and the normative theory of committees', *Journal of Theoretical Politics* 3: 259--76.

Black, D. and Newing, R. A. (1951) *Committee Decisions with Complementary Valuation* Glasgow: William Hodge.

Cairncross, (Sir) A. (1992) 'From theory to policy-making - economics as a profession', *Banco Nazionale del Lavoro Quarterly Review* 180: 3--20.

Cairncross, (Sir) A. (1993) Transcript of interview with I. McLean and A. McMillan, Oxford, 22.7.93.

Catlin, G. E. G. (1927) *The Science and Method of Politics* London: Kegan Paul.

Coase, R. H. (1937) 'The nature of the firm', *Economica* (new series) 4: 386--405.

Coase, R. H. (1981) 'Duncan Black: a biographical sketch'. In: G. Tullock (ed.), *Toward a Science of Politics: Papers in honor of Duncan Black* Blacksburg, VA: VPI Public Choice Center.

Coase, R. H. (1993) 'Duncan Black 1908--1991', *Proceedings of the British Academy* 82: 353--65.

Cohen, M. N. (ed.) (1979) *The Letters of Lewis Carroll* (2 vols continuously paginated) London: Macmillan.

Condorcet, M. J. A. N. de Caritat, marquis de (1785) *Essai sur l'application de l'analyse à la probabilité des décisions rendues à la pluralité des voix* Paris: Imprimerie Royale.

Cornford, J. (1963) 'The transformation of Conservatism in the late nineteenth century', *Victorian Studies* 7: 35--66.

Cox, G. (1987) *The Efficient Secret: the Cabinet and the Development of Political Parties in Victorian England* Cambridge: Cambridge University Press.

Cox, G. (1991) 'SNTV and d'Hondt are "equivalent"', *Electoral Studies* 10: 118--32.

Cox, G. (1993) 'Strategic voting equilibria under the single non-transferable vote', paper for APSA, Washington, DC, September.

Cox, G. and McCubbins, M. (1993) *Legislative Leviathan: party government in the House* Berkeley and Los Angeles: University of California Press.

Cox, G. and Niou, E. (1994) 'Seat bonuses under the Single Nontransferable Vote system: evidence from Japan and Taiwan', *Comparative Politics* 26: 221--36.

Cox, G. and Rosenbluth. F. (1993) 'The electoral fortunes of legislative factions in Japan', *American Political Science Review* 87: 577--89.

Cox, G. and Rosenbluth, F. (1994) 'Reducing nomination errors: factional competition and party strategy in Japan', *Electoral Studies* 13: 4--16.

Craig, F. W. S. (ed.) (1977) *British Parliamentary Election Results 1832-1885* London: Macmillan.

D'Hondt, V. (1885) *Exposé du système pratique de représentation proportionelle adopté par le comité de l'Association Réformiste Belge* Ghent: Association Réformiste Belge.

Dodgson, C. L. (1873) 'A discussion of the various methods of procedure in conducting elections' (Oxford: privately printed). In Black (1958), pp. 214--22.

Dodgson, C. L. (1874) 'Suggestions as to the best method of taking votes, where more than two issues are to be voted on' (Oxford: privately printed). In Black (1958) pp. 222--4.

Dodgson, C. L. (1876) 'A method of taking votes on more than two issues' (Oxford: privately printed). In Black (1958) pp. 224--34.

Dodgson, C. L. (1883) 'Lawn Tennis tournaments', letters to the *St James' Gazette*, reprinted in A. Woollcott (1939).

Dodgson, C. L. (1884) *The Principles of Parliamentary Representation* London: Harrison & Sons.

Dodgson, C. L. (1885) 'Supplement' and 'Postscript to Supplement' to *The Principles of Parliamentary Representation* London: Harrison & Sons.

Downing, P. B., and Stafford, E. A. (1981) 'Citations as an indicator of classic works and major contributions in social choice', *Public Choice* 37: 219--30.

Downs, A. (1957) *An Economic Theory of Democracy* New York: Harper & Row.

Droop, H. R. (1869) On the political and social effects of different methods of electing representatives, *Papers Read before the Juridical Society* 3: 469--507.

Droop. H. R. (1881) 'On methods of electing representatives', *Journal of the Statistical Society* 44: 141--202.

Dummett, M. (1984) *Voting Procedures* Oxford: Clarendon Press.

Farquharson, R. (1969) *Theory of Voting* Oxford: Blackwell.

Follett, M. P. (1918) *The New State: Group Organisation the Solution of Popular Government* New York: Longmans Green & Co.

Gardner, M. (1960) *The Annotated Alice* London: Anthony Blond.

Gardner, M. (1962) *The Annotated Snark* New York: Simon & Schuster.

Gardner, M. (1975) *The Annotated Alice* (2nd edn.) London: Nelson.

Gardner, M. (1983) *Logic Machines and Diagrams* (2nd edn.) Brighton: Harvester Press.

Gardner, M. (1990) *More Annotated Alice* New York: Random House.

Goldburg, C. B. (1994) 'The accuracy of game theory predictions for political behavior: cumulative voting in Illinois revisited', *Journal of Politics* 56: 885--900.

Goodacre, S. H. (1984) *Feeding the Mind: a Centenary Celebration of Lewis Carroll's visit to Alfreton in 1884* Burton on Trent: Parker & Son for the author.

Green, R. L. ed. (1953) *The Diaries of Lewis Carroll* (2 vols. continuously paginated) London: Cassell.

Grofman, B. (1981). 'The theory of committees and elections: the legacy of Duncan Black' in Tullock, G. (ed.) *Toward a Science of Politics: Papers in honor of Duncan Black* (Blacksburg, VA: VPI Public Choice Center), pp. 11--57.

Hare, T. (1859) *The Election of Representatives, Parliamentary and Municipal: A Treatise* London: Longmans Green.

Hare, T. (1873) *The Election of Representatives, Parliamentary and Municipal: A Treatise* (4th ed.) London: Longmans Green.

Huntington, E.V. (1938) 'A paradox in the scoring of competing teams', *Science* 88: 287--8.

Kendall, M. G., and Stuart, A. (1950) 'The law of cubic proportion in election results', *British Journal of Sociology* 1: 183--97.

Knuth, D. E. (1973) *The Art of Computer Programming* Reading, MA: Addison-Wesley, vol. 3.

Lijphart, A. and Gibberd, R. (1977) 'Thresholds and payoffs in list systems of proportional representation', *European Journal of Political Research* 5: 219--44.

Lijphart, A., Lopez Pintor, R., and Sone, Y (1986) 'The Limited Vote and the Single Nontransferable Vote: lessons from the Japanese and Spanish examples' in B. Grofman and A. Lijphart (eds.) *Electoral laws and their political consequences* (New York: Agathon Press) pp. 154--69.

Lubbock, Sir J. (1885) *Representation* London: Swan Sonnenschein.

Lytton, R. (1864) 'Report on the election of representatives for the Rigsraad' in *Great Britain: Parliamentary Papers 1864 vol. LXI* pp. 596--9.

McKelvey, R. D. (1979) 'General conditions for global intransitivities in formal voting models', *Econometrica* 47: 1085--111.

McLean, I. (1991) 'Forms of representation and systems of voting' in D. Held (ed.) *Political Theory Today* (Cambridge: Polity Press) pp. 172--96.

McLean, I. (1992) 'Rational choice and the Victorian voter', *Political Studies* 40: 496--515.

McLean, I. (1995) 'Independence of irrelevant alternatives before Arrow', forthcoming in *Mathematical Social Sciences.*

McLean, I. and London, J. (1990) 'The Borda and Condorcet principles: three medieval applications', *Social Choice and Welfare* 7: 99--108.

McLean, I. and Mortimore, R. (1992) 'Apportionment and the Boundary Commission for England', *Electoral Studies* 11: 293--309.

McLean, I. and Urken, A. B. (ed. and trans.) (1995) *Classics of Social Choice* Ann Arbor: University of Michigan Press.

[McMillan, A.] (1994) *Catalogue of the Duncan Black Archive at Glasgow University* Glasgow: Uiniversity Archives and Business Records Centre.

Marshall, J. G. (1853) *Minorities and Majorities: their relative rights. A letter to the Lord John Russell, MP on parliamentary reform* London: James Ridgway.

May, K. O. (1952) 'A set of independent, necessary, and sufficient conditions for simple majority decision', *Econometrica* 20: 680--4.

Miller, N. R. (1980) 'A new solution set for tournaments and majority voting: further graph-theoretical approaches to the theory of voting', *American Journal of Political Science* 24: 68--96.

Mitchell, W. C. (1993) 'Prolegomenon to public choice: the aborted work of George E. G. Catlin', *Public Choice* 77: 445--53.

Moore, H. F. [1920] *The History of Cumulative Voting and Minority Representation in Illinois, 1870-1919* (revised edition) Urbana: University of Illinois.

Moulin, H. (1988) *Axioms of Cooperative Decision Making* Cambridge: Cambridge University Press.

Mueller, D. C. (1989) *Public Choice II* Cambridge: Cambridge University Press.

Nanson, E. J. [1882] (1907) Methods of election. Paper read to Royal Society of Victoria on 12 October 1882. Printed in *Reports . . . respecting the application of the principle of proportional representation to public elections*, Cd. 3501, pp. 123--41. London: HMSO.

Page, N. (1983) *A. E. Housman: a critical biography* London: Macmillan.

Saari, D. G. (1994) *Geometry of Voting* Heidelberg: Springer-Verlag.

Sainte-Laguë, A. (1910) 'La représentation et la méthode des moindres carrées', *Comptes-rendus hébdomadaires de l'Académie des sciences* [de Paris] 151: 377--8. English translation in Lijphart and Gibberd (1977).

Salisbury, 3rd Marquess of (1884) 'The value of redistribution: a note on electoral statistics', *National Review* 4: 145--62.

Sawyer, J. and MacRae, D. (1962), 'Game theory and cumulative voting in Illinois: 1902-1954,' *American Political Science Review* 61: 936--46.

Sen, A. K. (1970) *Collective Choice and Social Welfare* Edinburgh: Oliver & Boyd.

Seneta. E. (1984) 'Lewis Carroll as a probabilist and mathematician', *Mathematical Scientist* 9: 79--94.

Seneta, E. (1993) 'Lewis Carroll's "Pillow Problems": on the 1993 centenary', *Statistical Science* 8: 180--6.

Taagepera, R., and Shugart, M. S. (1989) *Seats and Votes: the effects and determinants of electoral systems* New Haven: Yale University Press.

Tullock, G. (ed.) (1981) *Toward a Science of Politics: Papers in honor of Duncan Black* Blacksburg, VA: VPI Public Choice Center.

White, A. K. (1945) *The Character of British Democracy* Glasgow: Craig & Wilson.

Woollcott, A. (ed.) (1939) *The Complete Works of Lewis Carroll* London: Nonesuch Press.

Part 1. The Life and Logic of Lewis Carroll

The external events of Lewis Carroll's life were simple. His real name was Charles Lutwidge Dodgson, and he was a third child, being the eldest son, in a family of seven girls and four boys. On both sides, the family was well-connected and long settled in Northumberland, where they had scores of cousins. His father, a Church of England clergyman, had had a distinguished career at Christ Church, Oxford, and, when Carroll was twelve years old, accepted a living at Croft in Yorkshire, where the family circumstances were those of comparative opulence. To begin with Carroll had been educated at home by his father. With the change of residence he was sent to a private school ten miles away, which, in the preceding generation, had produced a number of notable scholars[1]; and, living within reach of his parents and sisters, Carroll was fairly happy. The first report of his 'kind old headmaster', as he was afterwards to call him, was so perspicuous that it deserves the quotation it has often been given.

> He possesses, along with other and excellent endowments, a very uncommon share of genius. Gentle and cheerful in his intercourse with others, playful and ready in conversation, he is capable of acquirements and knowledge far beyond his years, while his reason is so clear and so jealous of error, that he will not rest satisfied without a most exact solution of whatever appears to him obscure. He has passed an excellent examination just now in mathematics, exhibiting at times an illustration of that love of precise argument, which seems to him natural.
>
> I must not omit to set off against these great advantages one or two faults, of which the removal as soon as possible is desirable, tho' I am prepared to find it a work of time. As you are well aware, our young friend, while jealous of error, as I said above, where important faith or principles are concerned, is exceedingly lenient towards lesser frailties - and, whether in reading aloud or metrical composition, frequently sets at nought the notions of Vergil or Ovid as to syllabic quantity. He is moreover

[1] Cf. Sophia E De Morgan, *Memoir of Augustus de Morgan*, (London, Longman, Green, 1882), p. 139.

> marvellously ingenious in replacing the ordinary inflexions of nouns and verbs, as detailed in our grammars, by more exact analogies, or convenient forms of his own devising. This source of fault will in due time exhaust itself, though flowing freely at present ... You may fairly anticipate for him a bright career.[2]

From this school he went up to Rugby for three years, but although he made good progress in his studies, he was so unhappy in the unduly male atmosphere, out of reach of his sisters, that he was allowed to return home; and at this stage he edited and wrote the family magazines.

His career at Christ Church was no better than sound, until he reached the final examination in Mathematics, when he was placed at the top of the list. He became Mathematical Lecturer at the age of twenty-three, and a few years afterward a Student, that is, a Fellow of his College; and his entire life, from the age of nineteen, apart from short holidays at favourite sea-side resorts and one trip abroad, was spent within the walls of Christ Church.

His sisters frequently visited him at Christ Church and two of his brothers followed him there in their studies. He had a deep love for his sisters and found all girl children likeable, provided they were pretty and socially acceptable: he had, from childhood, a distaste for his brothers which he concealed fairly successfully, a child who was a boy could gain his interest only if he had some redeeming feature such as being the son of a famous poet or painter. They were an exceedingly compact family, and none of them married until both parents had died. When the unmarried sisters went to live at Guildford, Carroll helped them to find a house and he always spent part of his vacations with them.

Carroll had a great affection for his College. He had been fostered in the ambition that, like his father, he would make his mark there; and this affection for Christ Church survived whatever trials it met.

He was totally unsuccessful as a tutor. Partly this was because he was a poor mathematician [by conventional standards]. The subjects he taught were Euclid, algebra, and arithmetic, at the level of a first-year undergraduate course today; and his attempts to do serious Mathematics were feeble. His book on Determinants, published quite early, a sort of apprentice piece, showed an interest in Logic rather than Mathematics, though admittedly the two are difficult to separate. His work, at a later date, on Euclid's parallel, appears to have been unimportant, and he was not aware of the existing developments in Geometry being made by writers on the Continent and even in England. On the side of analysis, he failed to master the Calculus. He read much in general literature, and little else; when he knew that others had worked on a problem, in Mathematics or Logic, his interest flagged: he preferred to take up some new problem, and, working at it in bouts of intense concentration, arrive where his own thought would lead him. Considering that

---

2 Stuart Dodgson Collingwood, *The Life and Letters of Lewis Carroll (Rev. C L Dodgson)* (London, T F Unwin, 1899), p. 25.

Carroll was a poor mathematician, without an interest in his students, we can give some credence to the story, that on one occasion the members of his class signed a round robin and presented it to the Dean, asking that they be given another tutor in Mathematics.

When Carroll gave up the struggle to teach Mathematics and resigned from his Mathematical Lectureship at the age of fifty, however, he immediately sought the opportunity to teach Logic in the girl's high schools near Oxford. He liked his students, he enjoyed the work, and some of the girls afterwards spoke with enthusiasm of his teaching.

He was scarcely the man to be a success in his own college. In the early years he had a few close friends, but by the time he attained middle age, the number of his friends had dwindled almost to vanishing point.

> But I was thinking of a plan
> To dye one's whiskers green,
> And always use so large a fan
> That they could not be seen.[3]

Carroll was different from others and difficult to know: we cannot be sure that, had they known him better, his colleagues would have esteemed him more highly.

In middle age his child friends became more and more numerous. They visited him at Oxford and he found lodgings for them just outside the college gates and entertained them in his Christ Church rooms, giving rise to some concern among his colleagues. He would borrow a child from her parents and take her down to his boarding house at the sea-side: the company of children or of young girls had become a necessity to his peace of mind. As he got older he continued these friendships until the girls had got well into their teens, and took a pride in 'defying Mrs. Grundy', as he saw it, and indulged himself in kissing them.

But this was long afterwards, so let us return to Carroll at the time when he was a young Mathematics don, with, to all appearance, a good orthodox career ahead of him. A new Dean, H G Liddell, had just been appointed and had brought to Christ Church his handsome wife, Spanish in appearance, and his growing family of beautiful children.

Carroll was captivated by them. In his own home he had devoted himself to amusing his sisters, telling stories, acting plays with his marionettes, devising puzzles and writing verses. He was the perfect companion for a child and he was able to love the Liddell children directly and with an absence of concern. Then came the time of excursions to the river, croquet in the Deanery garden; games and puzzles and chess on wet afternoons, and always the story-telling.

---

[3] From 'Through the Looking-glass', reprinted in A Woollcott (ed.) *The Complete Works of Lewis Carroll* (London, Nonesuch Press, 1939), p. 225.

Isa Bowman, one of his child-friends, said in her biography 'He had a curiously womanish face ... there seemed to be little strength in it.'[4] The earliest photograph that we have of him, at the age of twenty-three, has 'a tinge of melancholy' in it, says Walter De la Mare[5]. To me these photographs are of a man of delicate feelings racked with cares and anxieties which are not of his own creation and from which he cannot escape. He had become foster parent to the numerous group at home, before he reached the years when he could have accepted these duties as normal and matter-of-fact. One side of his fate was the glory of telling stories, playing with marionettes, performing tricks of sleight of hand, devising puzzles, inventing games, writing verses: the other was a consuming concern for each of a number of individuals. The nexus of love and responsibility, joy and fear in which he had been caught, could not be shorn by an appointment at Christ Church and a betterment in his own affairs alone. He could gain freedom, if at all, only with isolation and with time.

The Liddell children he was able to love directly and with an absence of concern. With them came the time of excursions on the river, croquet in the deanery garden, games and puzzles and chess on wet afternoons, and always the story-telling: Carroll had served his apprenticeship at Croft and was the perfect companion.

On the fourth of July, 1862, there was the famous party on the river, with the three Liddell children, Carroll and his friend Canon Duckworth. Beginning with the white rabbit going down the hole, the stories were better than ever and Alice Liddell asked him to write them out, which Carroll promised to do. On the following day, travelling by train to London, he jotted down headings, and by October or November had completed the first draft of the story which he printed out neatly by hand, illustrated with his own drawings and presented to Alice Liddell. A facsimile edition of this first draft was later published under the title he had originally chosen, *Alice's Adventures Underground.* He added further episodes extending the story to almost twice the original length, had the illustrations done by Tenniel, the most famous of the artists working for *Punch*, and, three years to the day after the trip on the river, *Alice's Adventures in Wonderland* was published.

How the immortal story ever came to be written - apart from the external events leading up to it - is something of a mystery. Bishop Strong, who was a friend of Carroll's, spoke of him as 'an eccentric genius'[6]. It is difficult enough to know what goes on in the mind of the genius *or* the eccentric, and if we combine the two the task is well-nigh impossible - unless the genius relents and keeps a diary,[7] as Carroll did, from which we may get the occasional clue. He liked to have a record of everything that he was connected with, visits to theatres, the names of people that he

---

4 Isa Bowman, *The Story of Lewis Carroll* (London, J M Dent, 1899), p. 9.

5 [Walter De la Mare, *Lewis Carroll* (London, Faber and Faber, 1932), p. 27.]

6 [Quoted in Robert Phillips (ed.) *Aspects of Alice*, (London, Victor Gollancz, 1972), p. 46.]

7 Roger Lancelyn Green, *The Diaries of Lewis Carroll*, vols. i and ii (London, Cassell, 1953).

dined with, the titles of the pamphlets he was working at, and so forth. The notebooks in which he kept his diary in the years just before the writing of *Wonderland* have been lost, though fortunately we have the record for the period during which he wrote *Wonderland*[8].

Some years ago further documents became available, relating to this period in Carroll's life. I was interested in Carroll's writings on committees and went up to the Christ Church Library in the hope that it might yield some MS material; and when no MSS were available, I went along to the college Treasury, to consult the minutes of some of the meetings of the Governing Body. Occasionally I chatted with the Clerk of Accounts, and after a while, when we got talking about him, he mentioned that there were three letter books in an old cupboard, with copies of the letters that Carroll had written when he was Curator of the Common Room. Would I like to see them?[9]

The upshot of the matter was that a considerable number of documents came to light, most of them connected with Carroll's curatorship of the Common Room, with two notable exceptions. One was a quire of sheets forming a Letter Register, which gave a record of his correspondence, mainly with members of his own family, during the two years preceding the writing of *Wonderland.* This Letter Register had been of sentimental value to Carroll.

The other find was a large envelope containing cuttings from *Punch* and *The Times* and a number of reading lists he had made of articles from newspapers and journals. It gave some clue to his reading during the years 1855--62, just overlapping the period when he had begun to write *Wonderland.*

The item with a direct bearing on the *Alice* stories was the cutting of three letters to *The Times* of September 1862. At the International Exhibition in London, a frog had been put on display which was supposed to have been embedded in a block of coal, and a correspondent to *The Times* had suggested that this was impossible. Carroll's cuttings[10] take up the story from this point.

They are not particularly clever letters, but they came into Carroll's hands just when he may have finished the first draft of *Alice - Alice's Adventures Under Ground* - and was looking for further material to extend the story to publishable length.

---

8 We have the information for those years, 1859--1861, which Collingwood had incorporated in his *Life and Letters of Lewis Carroll.*

9 [For more details of these papers see Duncan Black, 'Discovery of Lewis Carroll Documents', *Notes and Queries*, Feb. 1953.]

10 The cuttings are in immaculate condition and could not have been displayed either at the Deanery or in the Common Room.

FROGS in COAL

TO THE EDITOR OF THE TIMES

Sir, - Your corespondent ..., in last Friday's paper, demands the expulsion from the International Exhibition of a frog said to have been found alive embedded in coal. He also accuses the exhibition of being an impostor, and attributes either credulity or inattention to the Commissioners, among whom he specially names Lord Granville. The only ground given for his conclusion is his own simple opinion that a frog could not have lived thousands of years so low down in the earth and having over it such an enormous weight as it must have had if what is said of it be true. It is in my power, however, to show that your correspondent's opinion is wrong, and I shall be obliged if you will allow me to do so.

1. My wife is prepared to state that she herself, may years ago, saw one of her father's workmen split open a piece of coal, and discover in the middle of it a moderate-sized frog or toad (she is not sure which) alive, and able to move, and she remembers distinctly the oval shape and smooth surface of the hollow where the animal had lain.

2. Samuel Goodwin, a stonemason, whom I myself have known these five-and-twenty years, and who is very trustworthy, states as follows: 'When I worked in the quarry at Kettlebrook, with Charles Aldridge, we sawed a stone through, about four feet thick, quite solid, and in the middle was a toad as big as my fist, and a whole about twice the size. We took it out, and it lived about half an hour, and then it died. We worked the stone, and it was used as a plinth stone in Birmingham Town-hall.'

I trust this will induce 'P.' to apologize to the exhibitor of the frog, as well as to the Commissioners.

Yours Respectfully,
JOHN SCOTT

Lilleshall Coal Depots, Great Western Railway,
Paddington, Sept. 17.

THE BATTLE OF THE FROGS

TO THE EDITOR OF THE TIMES

Sir, - I am surprised and grieved to find that any person still exists who is so sceptical as to doubt to possibility of the occurrence of living toads and frogs in solid blocks of coal or stone. Such a disbeliever would assuredly not be entitled to Christian burial, at least not in consecrated ground.

Permit me now to record in your widely-circulated columns two remarkable instances, which I doubt not, will immediately, and for ever, put an end to scepticism on this interesting subject.

Many years ago my grandmother, who was a most veracious and respectable old lady, informed me that one evening, as she was dozing in an armchair before the fire, she was startled by a loud noise. She instantly rose, and was somewhat terrified to observe a fine fat frog drop from her lap. A short time previously the servant had put a large lump of coal on the fire, and the sudden splitting of this into two or more pieces had produced the sound in question. She was perfectly convinced that the frog had been embedded in the lump of coal, though a coal-scuttle was close by her, from which it is just possible the reptile had escaped. However, I feel quite confident that my grandmother was right.

The second instance I present on the authority of Mr. Timothy Gosling, a highly respectable quarryman, and accurate observer of nature, whom I have known intimately during the last 30 years. Mr. Gosling was employed on the 1st of April, a few years ago, in blasting the rock which occurs near Birmingham, and is known as "Rowley rag". It was extensively used in paving the streets of that town. A shot fired a few minutes before noon brought down several large fragments of rock, and on breaking one of these with a sledge-hammer a toad suddenly appeared. Mr. Gosling was perfectly satisfied, and so am I, that the animal had been entombed in the solid stone, and liberated by the blow of the hammer. At first it seemed dull and stupid, having probably been stunned by the concussion. However, these symptoms did not last long; the creature became a great pet and survived several years. I should state that toads of precisely the same character had not infrequently been seen in this quarry. Now, Sir, geologists have the presumption to maintain that the 'Rowley rag' was formed by the agency of intense heat and ejected in a red or white-hot molten stream. But the fact of a toad having emerged from a solid block of this rock, in which not a crevice was detected, utterly disproves their absurd theory.

I should be sorry to expose myself to the charge of toadying any man on account of his noble birth, nevertheless, I cannot help expressing my opinion that the author of the scoffing letter in *The Times* concerning the "Frog in the Coal", at the Exhibition, deserves to be severely rebuked from presuming either to question the fact or sneer at the illustrious nobleman who presides over the Department of Science and Art.

I remain, Sir, yours respectfully,

Q.

London, Sept. 23.

TOAD IN COAL

TO THE EDITOR OF THE TIMES

Sir, — The controversy in your columns on the above subject reminds me of what I heard in Northumberland, as having occurred at Chillingham Castle, the seat of the Earl of Tankerville.

A slab of marble, forming one side of the chimneypiece, in either the dining-room or the drawing room, was observed to be always damp and somewhat discoloured, and partly from curiosity, and partly because the chimney was injured in appearance at that part, it was determined to examine the piece carefully. The slab was removed, and, I believe, was cut by a saw near the part where the unusual appearance existed, and a toad was discovered, alive, in the marble at this spot, and in the marble was found a recess of the size of the toad, and in which it exactly fitted.

I give you the story exactly as I heard it in the immediate neighbourhood of Chillingham Castle, and a single line from the Earl of Tankerville would confirm or disprove the statement, as its truth or want of foundation must be known to his Lordship, and to those resident thereabouts.

If the story is substantially true, I suppose that it is not more astonishing that a toad should be found in coal than in marble.

Your obedient servant,

GODFREY SINCLAIR

Ormsary, Lochgilphead, North Britain, Sept. 18.

When Carroll was extending the first draft of *Alice's Adventures Under Ground* so as to make it of publishable length, the only long insertion that he made was the episode of the Frog footman - and practically every feature of these letters to *The Times* got incorporated in *Alice in Wonderland.*

The frogs in the marble had waited for millennia and the Frog-Footman was prepared to wait indefinitely, 'for days and days'.

Furthermore Tenniel's illustration, done after Carroll's instruction, shows the Frog-Footman standing by a solid marble pillar and under a marble canopy.

There is no doubt that the provenance of the Frog-Footman is the letter to *The Times.*

His other 'Alice' book, *Through the Looking-glass* appeared half a dozen years later and collected up other stories Carroll had told to the Liddell children. We cannot open either of the *Alice* books and read for more than a page or two, without coming on some logical puzzle. We meet Logic dressed up in a bizarre and fantastic garb - Logic extending from the elements of the subject to the higher reaches, and even into parts of the subject where the solutions have not yet been found. The Logic of Lewis Carroll was just the Logic of anyone else, only a good deal sharper and applied in spheres to which most of us are strangers, ranging from Wonderland to wines.

Wonderland and *Looking-glass* are both dream stories: one episode slides into another in the manner we expect in dreams. If Alice is unable to cope with a situation she meets, she will wake up. She can continue the dream only by taking seriously the strange creatures she meets and only in she can follow their conversation sufficiently to keep them in play. They offer views she would not hear in ordinary conversation; she must gather *some* of the import of what they say, or they will disappear and her dream be snuffed out.

Invaluable to her in all this is her impeccable manner and her ability not to be startled, for the creatures in *Wonderland* and *Looking-glass* take up some unexpected topics.

Humpty-Dumpty, the most expert logician of them all, explains to Alice the nature of what has become known as the prescriptive or nominalist definition. And this is one of the most quoted passages in Carroll.

> 'When *I* use a word,' Humpty Dumpty said in rather a scornful tone, 'it means just what I choose it to mean - neither more nor less.'
> 'The question is,' said Alice, 'whether you *can* make words mean so many different things.'
> 'The question is,' said Humpty Dumpty, 'which is to be master - that's all.'[11]

Another instance in which Carroll takes up easily recognizable logical principle is in the exposition of Bishop Berkeley's philosophy given by Tweedledum and Tweedledee. Berkeley had argued that although we can be sure of the ideas in our own minds, we cannot argue from these ideas or images, to real objects lying behind them. The only things we can know are the ideas in our minds: the ideas are real, and that is all. I am entitled to say that my *idea* of an object exists: I am not entitled to say that the object exists. Berkeley gets rid of substance altogether, and a thing's being, consists in its being perceived or known. If we say that a thing exists, we mean that it is being perceived or thought by a mind.

Carroll develops this argument. Tweedledum and Tweedledee lead Alice to where the Red King was asleep under a tree.

> 'He's dreaming now,' said Tweedledee: 'and what do you think he's dreaming about?'
> Alice said 'Nobody can guess that.'
> 'Why, about *you*!' Tweedledee exclaimed, clapping his hands triumphantly. 'And if he left off dreaming about you, where do you suppose you'd be?'
> 'Where I am now, of course,' said Alice.
> 'Not you!' Tweedledee retorted contemptuously. 'You'd be nowhere. Why, you're only a sort of thing in his dream!'

---

11 [From 'Through the Looking-glass', *Complete Works*, p. 196.]

> 'If that there King was to wake,' added Tweedledum, 'you'd go out - bang! - just like a candle!'
> 'I shouldn't!' Alice exclaimed indignantly. 'Besides, if *I'm* only a sort of thing in his dream, what are *you*, I should like to know?'
> 'Ditto,' said Tweedledum.
> 'Ditto, ditto!' cried Tweedledee.
> He shouted this so loud that Alice couldn't help saying, 'Hush! You'll be waking him, I'm afraid, if you make so much noise.'
> 'Well, it is no use *your* talking about waking him,' said Tweedledum, 'when you are only one of the things in his dream. You know very well you're not real.'
> 'I *am* real!' said Alice and began to cry.
> 'You won't make yourself a bit realler by crying,' Tweedledee remarked: 'there's nothing to cry about.'
> 'If I wasn't real,' Alice said - half-laughing through her tears, it all seemed so ridiculous - 'I shouldn't be able to cry.'
> 'I hope you don't suppose those are real tears?' Tweedledum interrupted in a tone of great contempt.'[12]

Grant the premise that Alice exists only as a thing in the King's mind, and she can laugh or cry, but it won't make her any more real. Bishop Berkeley is not to be refuted in that way.

At other times the principle of Logic that Carroll has in mind is difficult to identify. Take this example where the White King waits with Alice for the arrival of one of the messengers.

> 'Just look along the road,' (says the White King), 'and tell me if you can see either of them.'
> 'I see nobody on the road,' said Alice.
> 'I only wish *I* had such eyes,' the King remarked in a fretful tone. 'To be able to see Nobody! And at that distance too!'

When the messenger arrives the King asks:

> 'Who did you pass on the road?' the King went on, holding out his hand to the Messenger for some hay.
> 'Nobody,' said the Messenger.
> 'Quite right,' said the King 'This young lady saw him too. Nobody walks slower than you.'
> 'I do my best,' the Messenger said in a sullen tone. 'I'm sure nobody walks much faster than I do!'
> 'He ca'n't do that,' said the King, 'or else he'd have been here first.'[13]

---

12 [From 'Through the Looking-glass', *Complete Works*, pp. 173--4.]

Here Carroll is concerned with the theory of universals. If there is yellow in my tie and yellow in the walls of the room, are they both instances of the same universal yellow? Does the universal yellow exist?

Or, take the problem in another form. If we can speak meaningfully of the numbers 3 and 7, does this imply that the numbers 3 and 7 must have some kind of existence? Bertrand Russell at one time believed that numbers must exist in some sphere or other.

For those who hold this view, Carroll puts the problem in the most difficult way: If we can speak meaningfully about Nobody, does this imply that Nobody must exist in some sphere? And if we say Yes, what do we mean by Nobody existing? Carroll pushes the view to the logical extreme.

There is repeated amusement over the meanings of words and over idiomatic phrases. When Alice speaks of *answering* the door, the Frog gardener enquires 'What's it been asking of?' She cannot explain herself, because she is *not* herself today. When she asks why she must do something *at once*, the reply is because she cannot do it *at twice*. The White Knight explains that his helmet had got stuck fast, 'as fast as lightning', though Alice knows that this is a different kind of fastness, and so on.

And so we might continue with these examples, but we have said enough already, I think, to show how vitally important Logic and Philosophy were for Carroll. Logical puzzles and logical principles were so much part of his mind, that he toyed with them and played with them even when telling stories to children. He had a mind which could not help being logical and he loathed inexactness of any kind.

Let me try to sum up in some way or other the position we have reached after this brief glance at his early life and at the *Alice* stories. Carroll was a person of extraordinary delicacy of feeling and even as a schoolboy, he had shown a passionate need for exact reasoning: in fact any kind of mistake in reasoning hurt him almost physically.

As an undergraduate at Christ Church he had studied Mathematics and Philosophy; and the principles of Logic and Philosophy continued to interest him so deeply that he incorporated them in his stories to the Liddell children and in *Alice's Adventures in Wonderland*. For Carroll a logical principle was more real than a table or a chair.

This was the man who, in his early 40's, during the years 1873--6, wrote three pamphlets on the Theory of Committees and Elections, but did not publish them: he intended to write a book on the subject at some later date. These pamphlets rank as one of the two most distinguished contributions to the Theory of Committees and Elections that have ever been made. They have come into prominence recently since a Theory of Elections and Committees - worked out in the exact terms of

---

13 [From 'Through the Looking-glass', *Complete Works*, pp. 205--7.]

Mathematics and Symbolic Logic - has been made the basis for a new type of Political Science from which much may be hoped[14].

1.1. Government by Logic

The political arrangements of the post-war world seem to many of us topsy-turvy, when we have to run faster and faster to remain in the same place, and we often invoke the name of Lewis Carroll. Yet it is not commonly known that at two stages in his life, Carroll did seriously concern himself with political theory, describing as he saw it, the way things ought to be. In both these contributions he was working at the roots of things, carrying out fundamental research, and, oddly enough, in dealing with the field of government - a province where, so to speak, he has no business to be - he is, without doubt, most conspicuously successful.

In 1876 he set out in a short pamphlet a theory of the committee; and the committee is our basic instrument of government, whether we are dealing with the government of the church, or the university, or the tennis club. He pointed out ways in which the ordinary committee procedure is defective; and he suggested that where important decisions are concerned, a different type of procedure should be employed, to make sure that the members arrive at the right decision.

This is one aspect of the matter. More important, in this pamphlet Carroll presents, in terms of symbolic logic, the theory of the committee which stands at the basis of a new and significant type of political theory which has come into existence during the last ten years [i.e., since 1958].

The suggestion he made about committee procedure enabled Carroll's own college of Christ Church to arrive at an important decision about the architectural design of its new belfry; and that is of some interest by way of an historical event. But the real importance of his pamphlet, as we view it today, is that Carroll is the forerunner of a new school of Political Science.

A decade after he had written his pamphlet[15] on committees, Carroll published his *Principles of Parliamentary Representation.* In this he took up the problem, which electoral arrangements will make parliament most truly representative of the voters? What is the best size of constituency, a single-member,

---

14 The reader, particularly the English reader, who thinks this an exaggeration - for admittedly the English literature on Politics is almost unaffected by the new trend - should consult William H Riker, 'Voting and the Summation of Preferences, an Interpretative Bibliographical Review of Selected Developments during the Last Decade', *The American Political Science Review*, vol. IV, No. 4, December 1961, pp. 900--11.

15 For a complete list of Carroll's publications on Politics, see Duncan Black, *Theory of Committees and Elections* (Cambridge, Cambridge University Press, 1958), pp. 189--92. In this book I have reprinted (pp. 214--34) three of Carroll's pamphlets.

or a two-member, or a three-member constituency, and so on? And if we have a multi-member constituency, with, say, three seats, should we give the voter three votes or two votes or only one? In his booklet he dealt with the classic problem of getting your legislature to be the best possible reflection of the views of the voters.

As with his theory of the committee, he developed his theory of parliamentary representation by means of formal logic or simple algebra and arithmetic. [Here] I will try to convey something of the flavour of his theories and give some account of how it was that Carroll come to concern himself with the problems of politics.

If we were approaching any writer other than Carroll it would be fairly safe to take it that his views on Politics arose out of the electoral system of the time. It was a great period of political theorizing in England and saw the appearance of the works of Walter Bagehot, and John Stuart Mill. The Great Reform Bill of 1832 cleared away the rotten boroughs and extended the vote beyond the landed gentry to the merchant classes, But an abundance of anomalies remained and it was known that reform would continue. For the next half century, 'reform', unless the word were qualified and given specific reference, meant reform of the franchise. Disraeli's Act of 1867 gave the vote to workers in the towns; but discussion went on up and down the country until Gladstone completed the process in 1884 and 1885, giving the vote to the agricultural workers and redistributing seats.

At the time when Bagehot, Mill and Carroll were writing, the English political scene was varied and exciting. Sometimes 1,000 voters would elect a member to parliament, sometimes 50,000. Many of the workers had no vote. The bulk of the members in parliament were elected by two- and three-member constituencies. In the three-member constituencies, to add to the apparent anomalies, the electors were given only two votes and not three, and they usually felt that they were being robbed of their vote. The English political system was ideal in only one way: nothing could have been better calculated to stimulate political discussion and to stir up the beginnings of a Political Science.

It was in this era of intense interest in politics that Carroll wrote his pamphlet on committees and his booklet on parliamentary representation. He was himself deeply conservative, in politics as in all other matters, always with a hankering after the world that had been and was passing away. He abhorred change in his personal life and would go for a dozen years to the same room at the same boarding house at the same seaside resort. His religious views at the end of his life were those of his boyhood. He rallied to the defence of Euclid against the threat to alter some of the proofs and change the order of the propositions. He always wore the same style of dark suit and, in summer or winter, wore grey or black cotton gloves. Was it not therefore perfectly natural that Carroll, perplexed by the thought of political change, should write on Politics?

On the face of it this might be so. Yet a feature telling against this view is that for an educated Englishman living at these times, Carroll seems to have shown comparatively little interest in the national scene. And this at any rate leaves open the possibility that some other explanation exists for Carroll's writings on committees and parliamentary representation.

The real explanation, I think, is less what was happening at Westminster than what was happening at Christ Church. His theory of committees, which I will consider now, was a direct response to events in the government of his own college. Another part of the explanation is that sometimes, in his best writing, Carroll is himself in the dark as to what he is writing about. He said of *The Hunting of the Snark*, written at the same time as his pamphlet on committees:[16]

> I was walking on a hillside one bright summer day, when suddenly there came into my head one line of verse - one solitary line - 'For the Snark was a Boojum, you see'. I knew not what it meant, then; I know not what it means, now; but I wrote it down and, some time afterwards, the rest of the stanza occurred to me, that being its last line: and so by degrees, at odd moments during the next year or two, the rest of the poem pieced itself together, that being its last stanza. And since then, periodically I have received courteous letters from strangers, begging to know whether 'The Hunting of the Snark' is an allegory, or contains some hidden moral, or is a political satire: and for all such questions I have but one answer, 'I don't know'.

Herein lies another part of the explanation for his pamphlet on committees. A theory of committees is also a theory of value or a theory of knowledge. In writing about the committee, Carroll was at the same time writing about his own feelings: and very queer feelings they were.

His circumstances at the time when he wrote his pamphlets were highly peculiar and he was at his most neurotic. Alice Liddell was blooming into a fine young woman who would soon leave Christ Church and move outside Carroll's circle. Her engagement was announced to a Christ Church undergraduate, an aristocrat of her own age. Long afterwards Carroll published *Sylvie and Bruno*, which is an exceedingly dull novel, its scene sometimes in fairyland, sometimes in this world, about the two loves of a young woman as observed by an old man. At one stage the tone is almost lascivious and disturbing when the narrator, the old man, gloats over the young girl just bursting into life and expresses with some naiveté his feelings towards her.

Then a recent biographer noticed a significant feature of the story. The name of the heroine is Lady Muriel Orme. Now each summer the Liddells drove down in their coach-and-four to Llandudno, a seaside resort whose distinguishing feature is the limestone headland which goes out like a squat worm with its nose in the sea: the name of the headland is the Great Orme. Carroll had never visited the

---

16 Quoted by Alexander L Taylor, *The White Knight: A Study of C L Dodgson*, (Edinburgh, Oliver and Boyd, 1952), p. 154. Cf. also Collingwood, *The Life and Letters of Lewis Carroll*, p. 173, and Green, *Diaries*, vol. ii, pp. 351--2.

Liddells at their country house[17], but in his childhood he had spent a holiday a few miles away, within constant sight of the Great Orme. He would also be familiar with the portrait by Sir William Richmond of the three Liddell sisters on the Great Orme. Lady Muriel Orme, the heroine of his story, is one of the Liddell sisters.

As if this were not enough, the young man to whom, at the beginning of the story, Lady Muriel Orme is engaged, is named Eric Lynd. Now Alice Liddell's fiancee, and afterwards her husband, was Reginald Hargreaves of Lyndhurst, Hants, so that the identification is complete. It is Alice Liddell whose love affair Carroll is describing in *Sylvie and Bruno.* Towards the end of the story Lady Muriel Orme realises that she cannot esteem her fiancee, Eric Lynd, in the way she would need to for marriage, and she rejects him in favour of a man of sterling character, twenty years her senior, who is of course none other than our old friend Carroll, masquerading under another name.

Sylvie and Bruno is our clue to Carroll's state of mind at the time 1873--6 when he was composing his theory of committees. He was in love with Alice Liddell, beautiful, accomplished, aristocratic, twenty years younger than himself and soon to marry an undergraduate - a very able open-air type, a first-rate cricketer and interested in his estates in Hampshire. To complicate matters Carroll was aware, we may believe, that although he was in love with Alice Liddell, he had no wish for marriage.

He was subject to the most conflicting desires and suffered quite extraordinary degrees of perplexity; and for a time his personality was in some measure disrupted. This period passed, but while it lasted his personality was divided and unstable and he was subject to agony and torment. Some of his actions were foolish and, to outward appearance, almost vicious. He went out of his way to attack Dean Liddell and even the Liddell daughters; and as it happens the main incident is one which has an immediate link with his theory of committees.

Much of Liddell's ambition for Christ Church was to make architectural improvements in the already beautiful building; among other changes the old belfry had been demolished and replaced by a temporary wooden structure; and it was on this belfry, most of all, that Carroll's ire and fury became concentrated. He ridiculed the temporary wooden structure as 'the tea-chest' in a broadsheet which includes a passage of remarkable disrespect to the Liddell daughters. He pictures them in the guise of birds gathered round the pool in the middle of the great quadrangle of Christ Church, to see which fish they can capture. The birds favour the nobler kinds of fish - the aristocratic undergraduates - and chiefly the Gold-fish - that is, preferring the wealthy aristocrats; and some of the birds are even King-fishers - Prince Leopold, the son of Queen Victoria, was one of the undergraduates at the time.

[Here is] a picture of Dean Liddell as seen by Lewis Carroll, a picture that is all the truer from Carroll being himself unaware of what he was writing about. Dean Liddell is the Bellman in *The Hunting of the Snark.*

---

17 This seems virtually certain: cf. Green, *Diaries*, vol. i, pp. 168--172.

'Just the place for a Snark!' the Bellman cried,
As he landed his crew with care;
Supporting each man on the top of the tide
By a finger entwined in his hair.

'Just the place for a Snark! I have said it twice;
That alone should encourage the crew.
Just the place for a Snark! I have said it thrice;
What I tell you three times is true....

The Bellman himself they all praised to the skies -
Such a carriage, such ease and such grace!
Such solemnity, too! One could see he was wise,
The moment one looked in his face!...

He had brought a large map representing the sea,
Without the least vestige of land:
And the crew were much pleased when they found it to be
A map they could all understand.

"What's the good of Mercator's North Poles and Equators,
Tropics, Zones, and Meridian Lines?"
So the Bellman would cry; and the crew would reply
"They are merely conventional signs!

"Other maps are such shapes, with their islands and capes!
But we've got our brave Captain to thank"
(So the crew would protest) "that he's bought us the best -
A perfect and absolute blank!"

This was charming, no doubt; but they shortly found out
That the Captain they trusted so well
Had only one notion for crossing the ocean,
And that was to tingle his bell.'

Carroll himself in the poem was, of course, the Beaver.

'There was also a Beaver, that paced on the deck,
Or would sat making lace in the bow:
And had often (the Bellman said) saved them from wreck,
Though none of the sailors knew how.'[18]

---

[18] [From *The Hunting of the Snark*, reprinted in *Complete Works*, pp. 680 and 683.]

It was Liddell's alterations to the Christ Church belfry that led Carroll to the essential step in his theory of committees.

Carroll entered Christ Church at the time when it was just about to begin changing from a medieval into a modern institution. We are told that about 1830 'the Dean and Canons in rotation nominated young men to Studentships. In many cases these nominations were mere matters of favour, and were bestowed on the sons of Canons, or some other of their relations and friends, without regard to merit. But the Dean and some of the Canons had begun of late to nominate Commoners of the House on the recommendation of the Censors and Tutors'.[19] Until after the middle of the century there was little encouragement to academic work and the standing of the college in the university visibly declined.

The Dean under whose aegis the college had lost ground, Thomas Gaisford, was himself a fine scholar who enjoyed the goodwill and respect of his fellows, and Carroll regarded the man who had welcomed him to Christ Church with a certain amount of affection. Gaisford disdained all change, and when in 1850 a Royal Commission was established to inquire into the state of the university and colleges of Oxford and to recommend changes in their administration and studies, he ignored its communications and treated it as if it did not exist. Yet he was compelled to witness the first stirrings of the new age; and in 1855 the Dean and Canons were forced to give up their immemorial right of nomination to Studentships. Their parting shot was an aggrieved comment not without its element of truth, though scarcely relevant in their own defence, on the undesirability of making awards on 'mere intellectual merit'.

At this time, the middle of the century, the college was still administered, as three centuries earlier, by the Dean and the Cathedral Chapter. The new governmental organization for the college was created by an Ordinance of 1858 and by the Christ Church (Oxford) Act of 1867. Now all the main issues, educational policy and the management of the properties and revenues (except as regards the Cathedral) were brought under the control of a Governing Body which was to consist of the Dean, the Canons and the Students; and in it the Students would form the majority.

When Gaisford died his successor was, by contrast, one who had stood in the vanguard of reform and had in fact been a member of the Royal Commission. H G Liddell, joint-author of 'Liddell and Scott', the famous *Greek Lexicon*, and father of Alice, became Dean in 1855 and his reign extended to 1892. Now that the college had been aroused from its torpor there were bound to be extensive educational changes; in particular it was necessary to arrive at a new definition of the rights and duties of the Students who had been given a main voice in the formation of policy. A strong-natured man like Liddell would inevitably hasten and amplify these changes and make them more thorough-going. But, as it happened, Liddell

---

19 H L Thompson, *Henry George Liddell, a Memoir*, (London, John Murray, 1899), p. 15.

was to make his mark in his own generation and to leave visible evidence of it for later generations, by a trait of character which could scarcely have seemed of much account at the time of his election. As a young man he had conceived a deep interest in art and architecture, which might never have found any outlet more significant than the artistic 'doodling' that he accomplished during committee meetings; but once at Christ Church he found and made new opportunities for his artistic impulses and embarked without delay on a series of improvements and transformations of the college buildings. Starting with his own house, he converted the Deanery into a gracious place of residence; then the ancient cathedral was extensively improved and in the course of time the quadrangle and belfry were to change their appearance, Fell's Buildings were to be demolished and the new Meadow Buildings rise in their place.

One reason then why Carroll was attracted to the theory of committees (one of the subjects in which his abilities as a logician are demonstrated) was the changes that were taking place in the structure of government of Christ Church and the significance for the life of the college of the decisions that were being taken by the Governing Body. Every Student of Christ Church was vitally interested in these matters; and Carroll, it may be felt, approached the whole subject in the manner that was natural to a logician of his calibre. Yet this explanation, which might on the face of it seem sufficient, would, as it happens, miss out a quite essential element, for Carroll's contribution to the theory of committees was no less due to his state of mind at the time when he was writing. This can best be understood by pointing to the two motives, each largely of an unconscious nature, from which his work derives.

Christ Church reproduced the same formal pattern as his own family circle. It was a closed society occupying its own house - in fact, the college is always referred to by its members as *the House.*[20] It was his father's old college, and his feeling of esteem and loyalty had been awakened before ever he set eyes on it. At its head Dean Liddell swayed its policies with the same appearance of reasonableness and the same dominance as his father did at home, and the Liddell daughters were a refined and aristocratic version of his sisters.

Without any change towards the circle at home, the same attitude that Carroll had towards his own family came to be extended to Christ Church - an attitude of reverence and solicitude, in which he was always asking himself how he could help. His nature made it difficult for him to conform altogether to the general pattern of college life, but if we judge loyalty by the test of being driven inevitably to protect its interests, Christ Church has had no more loyal son. And we will find that at least his first two pamphlets on committees gave expression to the desire to protect and serve his college.

The other main motive or unconscious drive from which these pamphlets spring is of a less admirable nature. Here again we can only state our view rather baldly but the reader may be referred for a fuller consideration of the subject to two

[20] The rendering into English of its proper legal title, *Aedes Christi.*

biographies.[21] Carroll's relations with the Liddell family were highly peculiar and he had in some sense or other fallen in love with his old child-friend Alice Liddell, though without the normal desire to marry her, and he had come to feel the most intense animus against the Dean and Mrs Liddell.

The evidence that chiefly concerns us is to be found in some of the pamphlets which he published on Christ Church affairs at practically the same time as he was writing on elections and mainly in *The New Belfry* and *The Vision of the Three T's*. Before this he had expressed himself freely, mainly in humorous skits, on numerous matters relating to college and university policy. He had resigned the office of Public Examiner in Mathematics for the university as a protest against the change in regulations permitting science students to give up Classics after passing Moderations (*The New Examination Statute*, 1864); he had taken part in the Jowett controversy (*The New Method of Evaluation as Applied to* π, 1865) and had written humorously (1865) on Gladstone, the Member for Oxford University, and protested (1867) against converting the Parks into cricket fields. These and other broadsheets and pamphlets had all Carroll's fondness for play upon words and contained much good fun. But with the appearance of *The New Belfry* (June 1872) and *The Vision of the Three T's* (April 1873) there was a change to a different key.

With his general concern for Christ Church we can easily imagine that Carroll felt some anxieties, among other things, about the completed design of the college buildings after the many alterations that Liddell had in mind had been carried through; and it would not be altogether unexpected that a person of Carroll's conservative tendencies should dislike some of the changes that were being made in the ancient fabric. But in writing about the matter Carroll went beyond the reasonable bounds of polemics, and his humour in some of the Christ Church pamphlets of the period with which we are concerned is sometimes savage and repellent.[22]

A really astonishing example of this is where, in *The Vision of the Three T's*, he allows himself, when ostensibly dealing with college architecture, to speak of the Liddell daughters under the guise of birds, as 'King-fishers' with an eye for the 'Nobler kinds' (of undergraduates) and 'Gold-fish' (i.e. wealthy undergraduates). In a dialogue concerning the fish that may be got from the pool in the middle of the Great Quadrangle of Christ Church, an angler gives some explanations:

> I will say somewhat of the Nobler kinds, and chiefly of the Gold-fish, which is a species highly thought of, and much sought after in these parts, not only by men, but by divers birds, as for example the King-fishers: and note that wheresoever you shall see those birds assemble, and but few insects about,

---

21 Taylor, *The White Knight*, chapters VII and VIII, and Derek Hudson, *Lewis Carroll*, (London, Constable, 1954), chapters VII, X and XI.

22 See the comment of John Francis McDermott in his edition of *The Russian Journal and other selections from the works of Lewis Carroll*, (New York, E P Dutton, 1935), pp. 26--7.

> there shall you ever find the Gold-fish most lively and the richest in flavour; but wheresoever you perceive swarms of a certain gray fly, called the Dun-fly, there the Gold-fish are ever poorer in quality, and the King-fishers seldom seen.[23]

The ambitions of the Liddells for their daughters were well known, and Christ Church was the traditional place of education for the sons of many of the noble families. Prince Leopold, the youngest son of Queen Victoria, was also in residence at this time. Carroll's meaning is clear enough and is put beyond doubt by Falconer Madan, who had himself known the ageing Carroll and who remarks that the passage 'gave some offence at the Deanery'.[24] Some of the rest of the 'fun' of the other tracts of this period is in little better taste than the passage we have quoted.

Whatever the allowance we make for Oxford's fondness for pamphleteering at this time, it is certainly strange to find a college don writing in such disrespectful terms about the family of his Dean. We cannot here consider in detail the reasons for Carroll writing as he did, for this would require a fairly lengthy study; but we will simply state what appears to us to be the almost inescapable conclusion. It is that by about 1872, when Alice Liddell, a girl of outstanding beauty and charm, reached twenty, surrounded by some of the most eligible young bachelors in England, and Carroll reached forty without much reputation in Christ Church, he realized that Alice Liddell, who had meant so much to him, was slipping out of his life. Even while Alice was still a child Mrs Liddell had not concealed her distaste for Carroll: the Dean considered him an impractical person whose opinion on business matters had often been a hindrance and rarely a help; and for years before this he had been discouraged from calling at the Deanery. Carroll felt frustrated and humiliated.

His reaction amounted to the attempt to triumph over Liddell and lower the stature of the Dean. By his pen, the instrument of his genius, he would alter the direction of architectural policy, and college policy, the objects at the centre of the life which he still shared with the Liddells and the objects which conferred or took away college reputation. And only the most warped of judgements could have allowed him to speak disrespectfully of the Liddell daughters.

The disreputable part he was playing would have been impossible to Carroll unless he had believed his motives to be quite other than they really were. He was aware that he felt concern for Christ Church and that he disapproved of the Dean's architectural plans. But the emotional impetus behind these various pamphlets is altogether disproportionate to the college matters to which ostensibly they refer, and his exacerbated and ragged feelings had their source nearer the heart. The model child in his own family and the favourite of his father had been rejected in his new family and he was in revolt against Dean Liddell and all his works, whatever they

---

23 Stuart Dodgson Collingwood (ed.), *The Lewis Carroll Picture Book,* (London, T Fisher Unwin, 1899), pp. 122--3.

24 Williams, Sidney H, and Madan, F, *Lewis Carroll Handbook*, (London, Oxford University Press, 1931), p. 54.

might be. The years 1872--6 were for Carroll the years of turmoil. They were also the last years of his period of high artistic creation.

At the same time as he was writing cantankerously about Christ Church architecture (or really, as we have suggested, about Dean Liddell), Carroll commenced working on the theory of elections and committees; and we must deal in some detail with the circumstances affecting his main writings on the subject, namely, his first three pamphlets.

The first pamphlet, *A Discussion* ... (1873), was written quickly for the meeting of the Governing Body that was to make the first appointment to the Lee's Readership in Physics at Christ Church and also an election to a Senior Studentship; and its preface, dated 18 Dec. 1873, explains:

> The following paper has been written and printed in great haste, as it was only on the night of Friday the 12th that it occurred to me to investigate the subject, which proved to be much more complicated than I had expected.

[H]is college was about to make two important teaching appointments, and the enquiring logical side of Carroll's mind got busy. He knew there would be a keen contest for one of these posts and he asked himself Which is the best method of electing a candidate? Most of us who are lucky enough to get a university education will find ourselves at various stages in our lives on appointing committees; and the scheme Carroll proposed on this occasion, 'the method of marks', will sometimes work better them any other.

For the sake of simplicity, let us suppose that there are only three electors whom we may call James, William and Robert, and let us denote the four candidates in the election by A, B, C and D. Suppose that after considering the merits or the different candidates, James reaches the conclusion that, in his opinion, the candidate B is more suitable for the post than any other, the next-best candidate is C, the third-best is D, and the weakest is A; then we can denote James's attitude towards the candidates by his schedule of preferences, on which B is placed higher than C, C higher than D and A stands at the foot of the schedule (Fig. 1.1). Similarly we can write down the schedule of preferences for William and Robert.

James	William	Robert
B	D	A
C	A	B
D	B	C
A	C	D

Figure 1.1

To find which candidate has the strongest claim to election, Carroll says, let us assign marks to them. Let us give to each candidate the mark 0 for each lowest place he gets on a schedule, the mark 1 for each second-lowest place, the mark 2 for each third-lowest place, and so on. When we do this the candidate to choose will be the one who scores the highest aggregate mark. For the diagram we have drawn, A would score 5 marks, B 6 marks, C 3 marks, D 4 marks; and we ought in the circumstances elect or appoint the candidate B.

> Marks for a = 0 + 2 + 3 = 5
> Marks for b = 3 + 1 + 2 = 6
> Marks for c = 2 + 0 + 1 = 3
> Marks for d = 1 + 3 + 0 = 4

There was nothing in the least profound about this suggestion: it was a good sensible method of election which almost anyone might have hit on. The Governing Body took it up and used it in one of the appointments it was making.

Carroll had written down several groups of these schedules and totted up the marks for different candidates. This was child's play. But the logical dissecting side of his nature was always liable to start into action. Behind the apparently commonplace group of schedules, might there not lurk some logical puzzle?

In view of Carroll's preface and the minute of the Governing Body, it had seemed, even before the appearance of the *Diaries*, that Carroll's method of marks might have been used in the elections of that day; and when the present writer explained this view to Mr J O Urmson, secretary of the Governing Body, and Mr Geoffrey Bill, archivist of Christ Church, they instituted a search, and one of the first documents which emerged from a bureau which had traditionally been the property of the secretary of the Governing Body, was the piece of paper on which the voting had been noted on the day in question. On the one side this was marked 'Election of R E Baynes (Analysis of Votes), Dec. 18. 1873', and on the other side was a record of the voting. This showed that Carroll's method of marks had been used at the first stage in making the appointment to the Lee's Readership in Physics: but when two of the three candidates practically tied for first place - Baynes 47 marks and Becker 48 marks - their names had again been put to the meeting and this time Baynes (11 votes) defeated Becker (9 votes).[25]

Now although Carroll both in the preface to his pamphlet and in the *Diaries* speaks of the importance of the elections to be held that day, he also knew that a further item of business was the proposal to be put forward by his friend Vere Bayne, that a sub-committee should be appointed to deal with the building of the belfry, and this proposal was adopted.

> A Committee of eight (4 elected by a majority of those present, 2 nominated by the Dean, 2 by the proposer (Mr Bayne)) was appointed with full power to choose a plan for the completion of the Belfry and to report the estimate

---

[25] There is no record of Dodgson's method having been used in the election of Paget to the Senior Studentship.

to the Gov[erning] Body with a view to proceeding with the work. (Minute of the Governing Body, 18 Dec. 1873).

Carroll's pamphlet The New Belfry shows how keenly affected he was by this topic, and further evidence of his interest is afforded by his attendances at the meetings of the Governing Body. The record shows that for a period of several years after this date he might 'cut' other meetings, but was almost invariably present if the discussion was to be on architectural business (the belfry and the quadrangle) or Studentships.

It seems likely, therefore, that the stimulus to Carroll's first pamphlet was not only the impending elections, but also the important decision which he knew was to be taken that day in regard to the belfry, about which he had expressed opinions which must have been disagreeable in the extreme to Dean Liddell.

The explanation of the almost instantaneous adoption of Carroll's method of marks lies in the atmosphere in which the suggestion was put forward. For many years around this period Christ Church was facing a crisis: its finances had been seriously strained by its building policy, and it was open to question whether it could wisely undertake all the reconstruction on which it was engaged; resources had been drained away from the strictly educational side and not all the permitted number of Studentships had been filled; even apart from this the educational side was in process of being transformed from an archaic pattern to a modern one. Now it is common experience in any institution that interest is greatest when the future is most insecure, and that times of near-crisis evoke maximum discussion of the possibilities that lie open and of the plans that may be formulated. In this atmosphere any view and any plan will secure a hearing and will be judged on its merits. It was in such circumstances making for open-mindedness and a heightened critical faculty that Carroll (as we know from the *ipsissima verba* of his pamphlets)[26] was able to explain in the Common Room and elsewhere the views which he was putting forward. And if his audience was specially prepared by circumstances to consider seriously the views of Carroll or of any other, it happened also to be one which was unusually well qualified to do so in virtue of its high level of intellectual attainment, though any more exact pronouncement on this point must await the appearance of a history of Christ Church at this period.

Thus when he placed the first of his pamphlets in the hands of members of the Governing Body, although, as he himself says, he was 'no orator', and although he carried little weight in its councils, his scheme was adopted. It had been explained to his colleagues in the days before the meeting and had commended itself to a sufficient number of them in the Common Room.

A result of the use of his method must have been to draw Carroll's attention to the main weakness of it; using the method of marks a candidate had been selected who failed to get a simple majority when put against one of the other candidates.

---

26 *A Method* ... (1876), § 3. [See Black, *Theory of Committees and Elections*, pp. 227--230.]

This was an odd outcome and must have set Carroll thinking about the next development in his theory.

After the Christmas vacation the Belfry Committee began its meetings and its intentions were no doubt a frequent subject of conversation. We can piece together fairly well the history of this committee from a document which may be referred to as the 'Vere Bayne Memorandum Book'.[27] In this notebook at various dates between 4 February 1874 and 14 December 1891, and more particularly for the early period, Vere Bayne, the secretary of the Governing Body, made rough jottings of minutes for some of the ancillary committees to which also he acted as secretary - the Belfry Committee, the Quadrangle Committee and the Committee on the Election of Senior Students in some cases without Examination. Indeed, it may have been to keep together the complexities of the Belfry and Quadrangle Committees that Vere Bayne started his record.

Running through all the entries for the early months there is the note of difference of opinion and perplexity as to the choice of design for the belfry. The first entry in the 'Memorandum Book' is the copy of a letter to Sir George Gilbert Scott, the architect who had already been responsible for the improvements in the structure of Christ Church Cathedral:

> The Committee being dissatisfied with all suggestions hitherto made for completing the Belfry and feeling the great importance of having a larger choice of designs in a matter of so much consequence, have resolved to apply to a limited number of Architects to furnish Sketch Plans; and they hope that Sir G G Scott will not object to be included in the number of those invited to compete'

Eight architects were approached, of whom three agreed to compete, then five others were approached of whom two, including Mr George Bodley, entered the competition; and later Scott, who at first had declined, also agreed to submit plans. Soon Scott appeared before the committee to enquire 'in what point or points the previously submitted plans failed to satisfy?' - a question not easy to answer since different people wanted different things.

The question of the belfry was again to be raised at the stated general meeting to be held on 18 June, and the evidence is overwhelming that acute difference of opinion existed among the members, with agreement only on the need that a decision should be taken. It was as a sudden preparation for this next meeting of the Governing Body that Carroll wrote his second pamphlet, *Suggestions* (1874). In his preface he alludes somewhat breathlessly to the circumstances:

> In the immediate prospect of a meeting of the Governing Body, where matters may be debated of very great importance, on which various and

27 This document came to light along with those referred to in Black, 'Discovery of Lewis Carroll Documents', *op. cit.*.

> conflicting opinions are known to be held, I venture to offer a few suggestions as to the mode of taking votes.

At various times during the past six months his mind had turned to elections, among many other subjects: he had had glimpses of a wider theory, but had not worked it out systematically; and the impending meeting about the belfry and the divided opinion of the members forced the pamphlet from him before he was ready to write on the subject. It gives only a suggested procedure to arrive at a decision with no attempt to show *why* this procedure might be considered satisfactory.

His proposal is that, for a start, all of the issues before the Governing Body should be listed and the first-preference votes for each of them should be counted; and if any motion is the first preference of an absolute majority of the members it should be adopted. but even though no proposal is able to get a majority of first-preference votes, it may be that there is some motion which can get a majority against each of the others which have been put forward; and , if so, this motion will be a satisfactory decision.

The minute of the Governing Body of 18 June reads:

> These proposals were voted upon, all four at the same time, with this result:
> 1. To adopt Mr Jackson's Tower — 9 votes in favour
> 2. To take Mr Bodley's Gateway as a basis — 2 votes in favour
> 3. To ask Mr Bodley for a fresh design, wholly in stone — 7 votes in favour
> 4. To adopt Mr Deane's Arcade — 5 votes in favour
>
> As this gave no absolute majority a vote was then taken, Bodley *v.* Jackson; for Bodley 17, for Jackson 9. Bodley had also a slight majority over either Deane or Scott. And it was carried by 16 *v.* 4 to get Bodley and Deane to work together if possible in preparing a design.

A final vote was then taken as to proceeding at once, with Bodley and Deane in combination if possible, else with Bodley: 19 Aye *v.* 6 No.

The state of opinion in the Governing Body was somewhat like that shown in Fig. 1.2:

9	**2**	**7**	**5**
A	B	C	D
B	A	B	B
C	C	D	A
D	D	A	C

Figure 1.2

There are four proposals before the committee, denoted by A, B, C, and D, each proposal standing for a particular design of the belfry. The members of the committee fall into various groups; 9 of them hold the first schedule of preference and consider the design A better than the design B, B better than C, and so on; 2 members think the design B best, A next best and so on; another group of 7 think the design C best and a group of 5 consider the design D best. Opinion is very divided.

Carroll's proposal was in effect that the Governing Body draw up a complete voting table or voting matrix, to show the number of votes for and against each of the others in a vote. The voting matrix or voting table (Fig. 1.3) shows that in a vote between A and B, A will get 9 votes and B 14; in a vote between A and C, A will get 16 votes and C 7; if the proposal B is put against the proposal C, B will get 16 votes and C 7; against D, B will get 18 votes and D 5, and so on. Although B is the first choice of only 2 members, a majority of the members prefer it to any of the other proposals which have been put forward.

Voting Table or Matrix for Fig. 1.2

AGAINST

		A	B	C	D
FOR	**A**	0	(9,14)	(16,7)	(11,12)
	B	(14,9)	0	(16,7)	(18,5)
	C	(7,16)	(7,16)	0	(18,5)
	D	(12,11)	(5,18)	(5,18)	0

Figure 1.3

When the members of the Governing Body saw that, in the case of the belfry, something of this nature was true, they were satisfied that the design which we have denoted by B, was the one which, in a very real sense, a majority of them wanted, and they adopted it. Carroll had enabled them to arrive at a decision - that was one thing: and he does it in such a way as to have the members satisfied that they had reached the right decision, the best open to them in the circumstances.

Carroll's prognostication had been correct. No proposal could get a majority of first-preference votes, but nevertheless the proposal that Bodley be entrusted to produce a fresh design could claim to have the support of a majority against each of the other proposals which had been put forward.

The only other information we have about the meeting is that it lasted for five hours; and it would be interesting to know after what disputes Carroll's method was resorted to. He had performed a signal service to his college in enabling it, in circumstances where opinion was divided and each member was attached to his own favourite scheme and feeling ran high, to arrive at a decision of manifest reasonableness. He had also in a curious way placed himself above Liddell, temporarily, as it were, usurping the functions of the Dean: for while the Dean presided officially, Carroll had chosen the procedure that the Governing Body would follow. The relish of his triumph was that he owed his ascendancy to the logic of his ideas and to the appeal to the sense of justice of his fellows. Both his urges were satisfied, his very real urge to help his college and his still stronger impulse of hostility to the Dean.

Carroll now intended to write a book on elections. He must already have accumulated notes and examples on the subject and probably had some notions about the general lines which the theory would take. But it seems that the long vacation of 1874 did not provide time for this work.

The autumn term commenced with the alterations to the quadrangle in full swing and afforded the occasion for perplexity and irritation. He published in *The Pall Mall Gazette* a rash and outspoken criticism of the proposed erection of cloisters, which he knew was favoured by Liddell and some of his colleagues; and in a further letter obstinately defended his right to make public criticism of this kind, so long as it was done before a decision on the matter had been formally arrived at by the Governing Body.[28] In point of fact Carroll's first letter was a strong criticism not only of the proposed building at Christ Church but also of the past building policy and of its educational and financial policy. The tone of the letters is insolent and hostile.

In the meeting of the Governing Body of December 1874, he voiced his disapprobation of the alterations to the quadrangle which a number of members had in mind; and then during the year that follows we are without any evidence of his feelings towards either Christ Church's architecture or its Dean. The diary entry after the statutory meeting of the Governing Body at the end of 1875, however, shows that he still bears in mind his desire to work on elections.

> Dec.: 18. (Sat). Election of Senior Students. After the usual dispute as to the way in which votes should be taken (a most complicated problem, which I still hope to work out some day) we elected Dalton, of C.C.C., & Payne Smith, of Trinity.[29]

Early in the following February an event occurred which set up another emotional crisis for Carroll and was instrumental in leading him to write his third pamphlet. Max Müller, who had been resident at Oxford for twenty-eight years, during most of which he had held one or other of two chairs, latterly that in Comparative Philology, had for some time felt strongly that he should devote the remainder of his life and energies to his self-appointed task of translating and editing some of the sacred books of the East and that he should waste no further time in teaching, 'doing work which others could do as well, or even better; while I had to

---

[28] Green, *Diaries*, vol. II pp. 333--4, entries for 29 Oct. and 4 Nov. 1874, where Mr Green quotes part of the first of these letters to *The Pall Mall Gazette*.

[29] This entry and that for 9 Feb. 1876, do not appear in the published version of the *Diaries* and are given here by courtesy of Miss F Menella Dodgson.

leave work undone which I could do, and ought to do'.[30] After his intention became known he received the offer of a chair at Vienna which would have freed him from all lecturing duties and allowed him to devote himself entirely to the task on which he had set his mind. Thereupon, at the thought of losing a scholar who had won such universal esteem and whose name was so closely identified with the university, Oxford opinion was stirred and a plan was very hastily put forward to enable Max Müller to remain; he might be relieved of all his teaching duties and retain half his salary, while the other half was used to engage a successor; and Liddell was to lay this plan before Convocation.

In the same way as Carroll had disapproved of the architectural changes at Christ Church, so also he disapproved of this plan; and in both cases, without knowing it, his real hostility may have been towards the author or instrument; Dean Liddell, rather than towards the plan itself.

There is no preface to Carroll's third pamphlet, but its place can be taken by the three broadsheets which he published on the Max Müller affair a few days before the pamphlet was written. Just as before 'conflicting opinions' were held by members of the Governing Body in regard to the choice of a belfry, so now 'There seems to be good reason for believing that there is among Members of Convocation a widespread feeling of dissatisfaction with the proposed Decree as it stands';[31] and, as is made clear by *The Times*[32] report of the debate that took place in Convocation, there was in fact extreme diversity of opinion.

Carroll's own feelings were divided and he felt himself tugged in different directions, for this time action against Liddell entailed also action against a valued friend. He reassured himself and others that his motives were entirely praiseworthy: 'an unbroken friendship of years with the Professor makes me feel safe from the imputation of any personal motives of hostility';[33] and this was perfectly true so far as his feelings to Max Müller were concerned.

Yet a reference to architecture reminds one that Dean Liddell too may be concerned: 'It appears incredible, while the University is able to provide so lavishly for the claims of Natural Science, for architectural improvements, &c., that so small a matter as this [provision of a pension quite apart from the emoluments of the chair] should be beyond its power.'[34] Again soon after the dispute had arisen he wrote his 'Easter Greeting to Every Child who Loves "Alice"';[35] perhaps his latest act of revolt against Liddell had awakened recollections of the happier times, a dozen years

30 F Max Müller and Georgina A Müller, *The Life and Letters of the Right Honourable Friedrich Max Müller, edited by his Wife*, (London, Longmans, Green, 1902), vol. I, p. 500.

31 Second broadsheet on 'The Professorship of Comparative Philology'.

32 Of 16 Feb. 1876.

33 Second broadsheet on 'The Professorship of Comparative Philology'.

34 Third broadsheet on 'The Professorship of Comparative Philology'.

35 Reprinted in part in Florence Becker Lennon, *Lewis Carroll* (London, Cassell, 1947), pp. 92--3.

earlier, when amid trips on the river and visits to the Deanery his genius had been loosed and he had been inspired to write his immortal stories. Carroll's feelings were indeed mixed.

It was during this very same fortnight, while he was exercised over the proposed decree, that Carroll had again begun to work on his theory of elections.

> Feb.: 9. (W). G[overning] B[ody] Meeting. The question of a system for taking votes at an election was again postponed. Afterwards I arranged with Bayne that we should try to get information as to the rules adopted in the other Colleges. Lawrence's opinion, given to-day, that 'the greatest number of votes does not necessarily mean an absolute majority', gives us the means of a final settlement of the matter, when there is a 'cyclical majority' that will yield to no other remedy.[36]

Just eight days after the meeting of Convocation with its excitement and mixed feelings, Carroll's third pamphlet was written - the one which entitles him to a position in the theory of elections and committees only a little lower than that of Condorcet.

Carroll had been caught in the grip of the theory of elections and committees and his understanding of the subject was second only to that of Condorcet. After 1876 it was still his intention to write his book. But he lacked the systematic approach of Mathematics on which Condorcet had been able to lean. The contributions he had made were artistic rather than scientific, a curious amalgam of conscious logic and unconscious symbolism; and the struggles which he was symbolizing did not recur - for the main struggle was in his own inner nature and not in the Christ Church committees, and after 1876 this was rapidly changing.

His mind had been challenged to find the solution to a problem which was very similar in nature to the problem of the committee; and the concern he felt for the proper administration of Christ Church had got him not merely to think about the problem but to feel it in his nerves and sinews.

Two years later the same problem emerged on the national level: Gladstone introduced proposals for the much-needed reform of the electoral system. Now, with the Christ Church experience behind him, Carroll sought to get a scheme of representation that would both meet the needs of the country and his own inexorable demand for logic.

His mind was like one of these shampoos which are known as double reagents: the first application gives you a little lather and the second leaves your head a mass of foam. When he approached a logical problem he would get excited about it and reach some conclusions that would give him intellectual satisfaction for a time. Some time later his mind would recur to the problem with an intense concentration and he would carry the earlier results much further. [We discuss Carroll's work on proportional representation in Part II.]

---

36 Not in published version of *Diaries*, See footnote above.

[Further light on Carroll the man and the logician derives from] a later phase in his life, when, at the age of fifty, he became Curator of the Christ Church Common Room, an honorary post which he held for nine years.

During the early part of his Curatorship he was confronted with a range of problems that were altogether new to him: he had become an administrator and a business man: and he tried to make good his lack of experience by pitting his intellect and his logic against the problems encountered.

He had given up his Mathematical Lectureship with the intention of settling down to literary and scientific work, but he was aware that he was getting crotchety: grown-ups were becoming more and more of a trial to his patience. 'Utopia is a pleasant and a well-ordered country, and enjoys many blessings to which our little island is a stranger. Some of these must, no doubt, be eternally despaired of (for example, no one is ever bored at a Utopian dinner-party, or overcharged by a Utopian cab-driver)'.[37]

He took up his duties as Curator in a spirit of self-sacrifice. 'I accepted office with no light heart: there will be much trouble and thought needed to word it satisfactorily: but it will take me out of myself a little, and so may be a real good. My life was tending to become too much that of a selfish recluse.'[38]

The duties of the Curator, before which Carroll stood appalled, were no more than to preside over the Wine Committee and the annual general meeting, to take an interest in the physical and spiritual comfort of the other residents and help to make the guests of the House feel at ease. A steward looked after the business side and the day-to-day arrangements, the occasional private dinners given by members, and so on. A broad and genial common sense were needed in the Curator, and, granted this, he might draw pleasure from the intellectual conviviality of the Common Room.

Carroll knew that in his own case things would work out differently. Almost a stranger to wines, he now set out to acquire the whole mystery of cellarage. 'Surely any Curator, worthy of the mane would be found, if physically tested ..., to possess a density varying directly, and a gravity varying inversely, as the potency of the Port - if tested anatomically ..., to have the word WINE neatly emblazoned on his heart.'[39]

During the first year of office he wrote 800 letters enquiring about wines, enquiring about the best temperature, ventilation and dampness for the cellars, in his violet ink, and made inventories of the bins and racks, to be able to trace the movement of a single bottle from bin to bin and from the cellar to the Common Room. He bought in quantity to get the benefit of discount, and at low prices long before the wines would reach maturity. Since the members' tastes might change

---

37 Charles Dodgson, 'Purity of Elections', *The St. James's Gazette*, 4 May 1881.

38 Green, *Diaries*, vol. ii, entry for 8 Dec. 1882.

39 [From 'Three Years in a Curatorship', reprinted in *Complete Works*, p. 1066.]

between times, he would sometimes buy only if the merchant would agree to accept it back later and pay the Common Room 5 per cent. per annum for the length of time the wine had been stored.

Even when all precautions had been taken a wine might not mature as had been expected.

> About a year ago panic seized us. One of two bottles (of Mouton Claret) had turned out bad ...; and suddenly the cry arose "All is lost!": wild words, such as "It is past its prime!" "It is worth only three shillings a bottle!" hurtled in the air: the very constitution of the Cellar was affected for a time: symptoms of diminished circulation and of slight consumption showed themselves. The Curator trembled, but would not quit the gory field in such frantic haste, or give the order ... to empty the remaining bottles into Mercury (the pool in the middle of the Great Quadrangle) - thereby certainly demoralising, and probably destroying, its scaly inmates. "We are but amateurs," he said to himself: "and, though Christ Church Students, we are still fallible!'" And the experts reported that the wine would reach the zenith of its prime in about four years.[40]

In his choice of wines Carroll made use of his theory of elections. He gave a party in his rooms for the wine Committee, who tested the various wines and ranked them in order of merit From these individual ratings Carroll deduced their opinion acting as a committee.

Tastings

	S. Emilion /84	Listrac /84	Pauillac /84
Hassall	best - 3	last - 1	nearly = Emilion - 2
Sayers	last of the three - 1	second - 2	best - 3
Skene	best - 3	2	1
Curator	last - 1	2	3
	8	7	9

The table shows a record of one of these tastings. The members' opinions differ, but Carroll was satisfied that they regarded the Pauillac as best and the Listrac as next

40 [From 'Three Years in a Curatorship', reprinted in *Complete Works*, p. 1067.]

best. On one occasion, after getting them to judge between two bottles of port, he disclosed that they were the same wine.[41]

He discovered that the value of a bottle of liqueur might vary enormously from one liqueur to another, from 36 fluid oz. in the case of Chartreuse to only 20 for Maraschino. Whereupon he wrote to his favourite wine merchant, from whom he expected some indulgence, asking him to bottle some liqueurs for the Common Room in 8 oz. bottles, of the type used for medicine, with glass ridges marking every 2 ounces.

He then devised a liqueur measure which he wanted the Common Room butler to use in making his charges. It was a ruler, graduated in shillings and pence, which the butler would stand upright by the bottle, and read off to the nearest penny the amount served.

Carroll had never been popular with the college scouts, as the servants were called, less so than ever at this time, no doubt, when he had more to do with them. As healthy a comment as any had been provided in a satire on various members of the college by an undergraduate who got sent down for his pains. Carroll's attitude to the scouts is:

> Ah scouts! a pamphlet I will surely write
> Which with a serpent's tooth will keenly bite.
> Your perquisites, your pilferings I'll betray,
> And turn to hellish night your garish day.

After he has been imprisoned for a time in the Belfry, to think over his misdeeds, two of the scouts bear Carroll aloft and pitch him into the pond in the middle of the quadrangle, with the leave-taking:

> Full fathom five e'en now he lies.
> Of his bones are segments made.
> Those circles are that were his eyes.
> Nothing of him that doth fade
> But doth suffer a sea-change
> Into something queer and strange.
> Goldfish hourly ring his knell.
> Ding-dong
> Hark! now I hear them, ding, dong, bell:[42]

So far the account of the activities of the Curator has made only a remote approach to Logic: but with this background of economic activity, Logic could not be long delayed. Pricing lends itself to Logic; and when a dispute arose about the price of tea

41 Williams and Madan, *Lewis Carroll Handbook*, p.xvii.

42 From *Cakeless* by Rev. John Howe Jenkins, quoted in Hudson *Lewis Carroll*, p. 218.

in the Common Room. Carroll settled all these issues promptly by what today would be known as Cost Accounting.

He presented this table (Fig.1.4) showing the costs of all items of food provided by the Common Room. In the case of Dessert the receipts failed to cover the costs, while members paid twice as much for coffee as it cost the Common Room. He prepared a new scale of charges which would just cover the costs and leave 10 per cent. profit for the Common Room in each case. The justice of the arrangement was immediately seen and the new prices adopted.

Then he ventured with less success into an economic problem which has been the subject of dispute from the nineteenth century to the present day. In the twentieth century books have been sold at their net price, shown on the dust jacket; and a similar system was in vogue throughout the first half of the nineteenth century: but the booksellers were much against it and about 1850, on the plea that it was a monopolistic practice and in restraint of freedom of trade, the booksellers succeeded in having this sale at a fixed price (resale price maintenance) abolished.[43] After that a bookseller sold at whatever price he thought the book would fetch. He would add 5 per cent. or 50 per cent. to the wholesale price, according to his view of what the traffic would bear.

FOOD:- Expenditure and Receipts.

	sent to rooms	5 p.m. Tea	Desert in C.R.	Coffee in C.R.	Totals
	[£.s.d.	£.s.d.	£.s.d.	£.s.d.	£.s.d.]
Tea	2 0 0	3 0 0			5 0 0
Coffee	4 0 0			4 0 0	8 0 0
Sugar		0 10 0	1 5 0	0 15 0	2 10 0
Milk	2 0 0	3 5 0		5 10 0	10 15 0
Cake, Fruit etc.	20 0 0	2 5 0	31 15 0		54 0 0
Totals	28 0 0	9 0 0	33 0 0	10 5 0	80 5 0
Charged in bills	29 18 9	14 6 6	27 10 1	21 4 6	93 9 10

Figure 1.4

The *Alice* books had a wholesale price of 4 s. and an advertised price of 6 s. The bookseller would sell somewhere between these two figures, but very rarely would his profit be 2 s.

Now early in his curatorship Carroll was struck by the corresponding arrangement in the sale of liqueurs. Some were sold to the Common Room at a heavy discount, 25 per cent., below their advertised price, while others were sold at a

43 Cf. Marjorie Plant, *The English Book Trade: an economic history of the making and sale of books* (London, George Allen and Unwin, 1965), pp. 432--44.

discount as low as 5 per cent.. After considering the expenses of publication, he took the view that the bookseller's profit was too large - "monstrously" large - and the author's profit on the *Alice* books too small.[44] Then, following the system that obtained in regard to liqueurs , he decided that they should be offered to booksellers at a lower discount on the advertised price[45], and that, instead of 2 s., the bookseller's profit would be 1 s.

There was a hue and cry among the booksellers. A pamphlet which Carroll wrote for the better understanding of all parties to the dispute, failed to convince them that this was a satisfactory arrangement. The *Alice* books were boycotted and after watching sales decline for some years, Carroll was forced to give in to the booksellers.

The episode of the Curatorship stresses that Carroll's response to the problems he was confronted with was always to fall back on strict reasoning. There was a proper way of tackling any practical problem. If only his logic were acute enough, he felt, the problem could be met: and in this reliance on the application of Logic to practice, taking logic as a means by which to achieve practical ends, he was, I imagine, different from most professed logicians.

After the first couple of years in the Curatorship - when the cellars were properly organised, the bins well stacked, when coffee, tea and dessert were charged at their proper prices, as determined by cost accounting, and the rules governing the procedure of the Wine Committee had been properly drafted - Carroll breathed more freely and began to devote more time to academic pursuits. He was moving towards the conception of himself as a logician; and, increasingly as time went on, it was to formal Logic that he devoted the energies of his remaining years.

Carroll succeeded in getting classes at the Girls' High School and two of the women's colleges at Oxford to teach Logic. He printed a series of papers, with examples and diagrams, some intended for the use of these classes; and he entered into a long correspondence on disputed points, with J Cook Wilson, the Professor of

---

44' His trouble was, first, that booksellers had the lion's share of what the public paid for a book, and, secondly, that their system of underselling one another was preposterous, as indeed it was. He calculated that, from the sale of a 6 s. copy of *Alice*, the author received 1s. 2d., the publisher 10d. and the bookseller 2 s. His view was that the author should have 2 s. 2d., the publisher still 10d., and the bookseller be content with 1 s.' (Charles Morgan, *The House of Macmillan (1843 - 1943)*, (New York, Macmillan, 1943), p. 108.

45 Macmillan, who published for Carroll on commission, inserted the advertisement at the end of his books: 'In selling MR LEWIS CARROLL'S books to the Trade, Messrs. MACMILLAN AND CO. will abate 2d. for payment within six months, and 10 per cent. for cash. In selling them to the Public (for cash only) they will allow 10 per cent. discount.' A bookseller, making payment within six months, would pay 4 s. for a copy of *Alice*.

Logic at Oxford.[46] He published his *Game of Logic* in 1886 when he was approaching his middle fifties, and his *Symbolic Logic, Part I, Elementary* appeared in 1896, a little before his death; the other two parts of *Symbolic Logic*, which he had projected, were never published. He had prepared for the task of writing on Logic, with a quite unwonted assiduity, and could justifiably tell his publisher that he had been working on his second book 'for twenty years or more.'[47]

The title *Symbolic Logic* is off-putting to all except the intrepid; and with Carroll's books this would be a mistake. He intended his earlier book for children, his later one for young people to help them to think straight and to enable them to see whether a conclusion followed from the premises of was a fallacy; and his books are admirably adapted to this aim.

His treatment is simple and delightful. First of all, let us suppose that, instead of dealing with arguments and propositions, as in Logic, we are simply dealing with some genus or class of things, say the books in a library; and let us divide up these books into their various species. Some of them may be old; let us divide off those that are old from those that are not old. Or, in general, let us choose any attribute at all, - call it x - and let us separate off the books that are x from those that are not x - or x', as Carroll calls it; and let us put the books with the attribute x in the top half of the square (Fig 1.5), and those with the attribute x' in the lower half. If x means old, we have collected old books in the top half of the square, new books in the lower half. This is our first dichotomy.

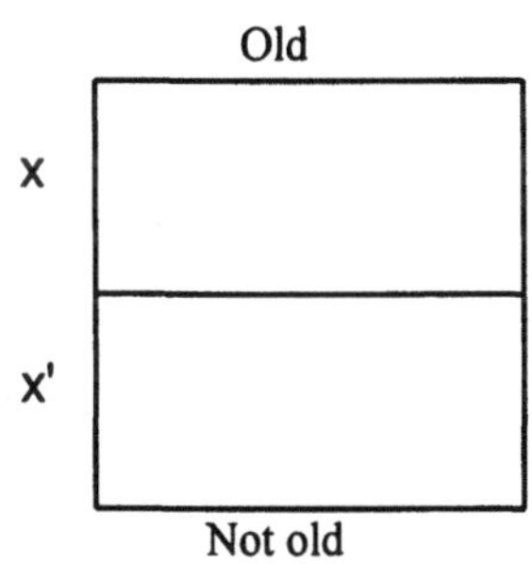

46 Cf. J Cook Wilson, *Statement and Inference* (Oxford, Oxford University Press, 1926), Vol. 1, p. xli--xliii.

47 Williams, Sidney Herbert, Madan, Falconer, and Green, Roger Lancelyn, *The Lewis Carroll Handbook*, (New York, Oxford University Press, 1962), p. 183. In 1855 he noted in his study of John Stuart Mill's *A System of Logic* (London, John W Parker, 1843) and in 1884 he was 'seeking a simpler notation' than that used by George Boole in *The Mathematical Analysis of Logic* (Cambridge, Macmillan, Barclay and Macmillan, 1847). cf. *Diaries*, vol. i, p. 44 and vol. ii, p.430. *The Lewis Carroll Handbook* lists his various papers on Logic, including his two contributions to *Mind*, and the editions of his two books.

Figure 1.5

Next let us choose another attribute, say English books (or books printed in English), and separate them off from books not printed in English. Let y.denote English and y' not-English, that is foreign. and let us place books with the attribute y on the left half of the square, those with the attribute y' on the right half of the square. This gives us our second dichotomy. The north-west cell now contains things with the attributes New and English, and the south-west cell things attributes Old and English, and so on.

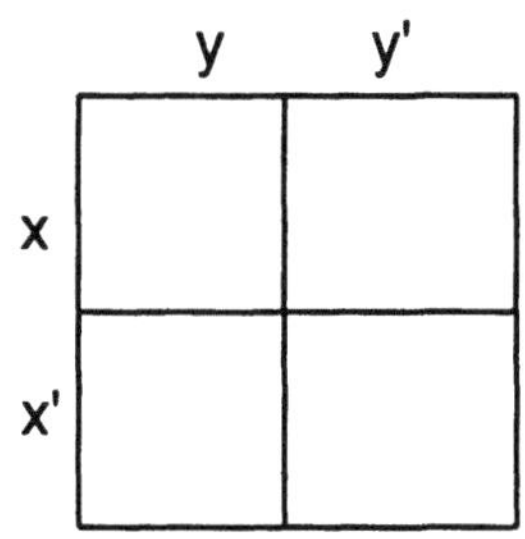

Figure 1.6

When there are one or more things in a cell, we can denote this by placing the figure 1 in it; when there are no things in a cell we denote this by placing the figure 0 in it.

To get a diagram that will be useful in Logic we must add a third attribute, dividing things up according as they have say, the attribute z or not z, that is z or z', for instance books that are bound and books that are unbound. Carroll deals with this by putting into the four inner cells we already have (Fig. 1.6), all the things with the attribute z, and putting all the things with the attribute z' into the outer border; and this gives us our completed diagram Fig. 1.7. We have now divided up all the things we are dealing with, according to the three attributes x, y, and z, getting in all $2^3 = 8$ classes of things, one corresponds to each cell in the diagram.

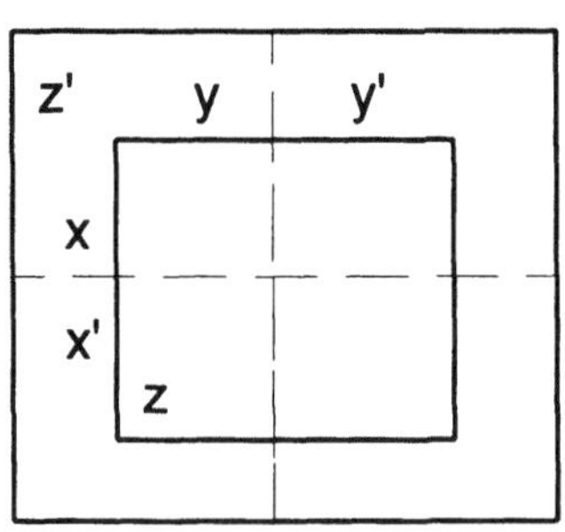

Figure 1.7

So far all we have got is an exceedingly neat way of classifying things. But Carroll goes on to show that this scheme of classification can be used to deal with all syllogistic reasoning. A syllogism consists of two premises, which must contain the same middle term, z or z'; and the conclusion is got by eliminating z or z'.

To deal with a syllogism all we need do is to replace the words and phrases in the premises by x, y and z, or x', y' and z', and employ the diagram. Doing this we can tell which conclusions, if any, follow logically from the premises.

For more complicated arguments, for example arguments with more than two premises, he again shows how to deduce the logical conclusions from the premises by means of diagrams. He shows how to draw the diagram for a four-fold, five-fold and up to an eight-fold division of the material: but of course the diagrams now become fairly complicated.

A diagrammatic treatment of this type, Carroll thought, was the best approach to Symbolic Logic for young people: it was easy for them to master, as his own experience showed; after a few lessons they could deal with the syllogism; and it led to a keener interest in abstract thought.

No one could have written more effectively for the audience he had in mind, and *Symbolic Logic, Part I, Elementary* quickly went through four editions. But it was not serious Logic - not Logic for logicians; and his books on the higher reaches of the subject never got published.

His diagram failed to achieve any vogue. A few years earlier the logician [John] Venn had introduced a similar type of diagram, using three intersecting circles instead of the squares. One can draw three circles almost without premeditation: their meaning is obvious at a glance; and it is the Venn diagram which is universally used in connection with the syllogism.

Again, to speak technically for a moment and cite Professor R B Braithwaite in his article on 'Lewis Carroll as Logician',[48] the view that Carroll had taken, that the subject term in a universal proposition must be regarded as existing, proved to be inconvenient; and again his work was by-passed by the logicians.

His book, even without emendation, remains admirable for young people. Its collection of imaginative examples has helped to brighten the pages of many a textbook. But it has played virtually no part in the development of logical theory.

By about 1950 I think you could say Carroll's books on Logic had reached their *nadir* in neglect, and from about then on the references to them in the journals have become more frequent. In 1951 C D Broad of Cambridge, ventured the opinion

---

[48] *The Mathematical Gazette*, vol. xvi, No. 219, July 1932, p. 175. The reader may also consult in connection with Carroll's work. L J Russell 'A Problem of Lewis Carroll', *Mind*, Vol. 60, No. 239, pp. 394--6, 1951; D G Brown, 'What the tortoise taught us', *Mind*, Vol. 63, No. 250, pp. 170--9, April 1954; Lennon, *Lewis Carroll*, chapter xv, pp. 264--284; and Martin Gardner, *Logic, Machines and Diagrams*, (Brighton, Harvester Press, 1983, first published 1958), pp. 45--51, 78 and 106.

that Carroll's diagrams were the neatest in Logic - and this is particularly true if you are working with 4 or more attributes.

Then in 1953 there was a sudden change in the picture. M[aurice] Karnaugh of the Bell Telephone Company published a new type of diagram - which was really the Carroll diagram subjected to a topographical deformation. And this Karnaugh Map is of very great use both in Boolean Algebra and in the Theory and Design of Electric circuits. It is mentioned, I believe, in most of the recent books on Boolean Algebra and Electrical Engineering.

[Karnaugh's] 2 attribute map [is the] same as Carroll's.

[For] 3 attributes, $2^3 = 8$cells (Fig 1.8):

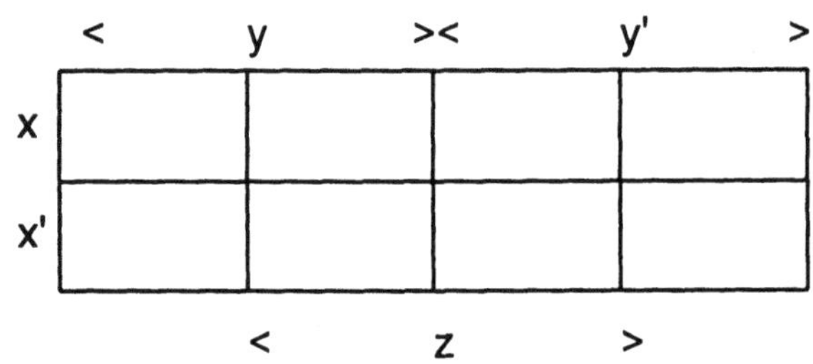

Figure 1.8

4 attributes. $2^4 = 16$ cells (Fig. 1.9):

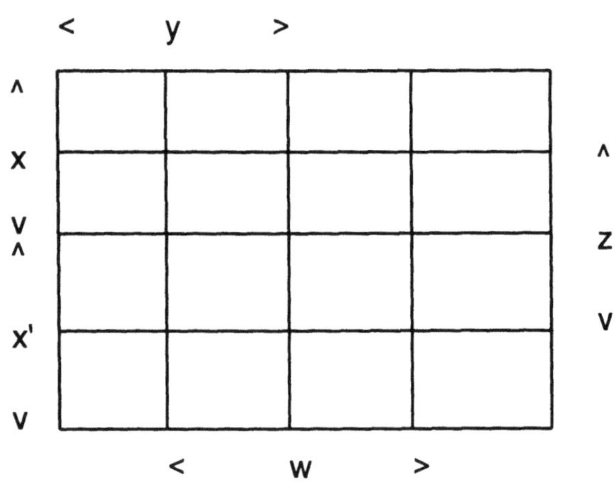

Figure 1.9

It will be found that, using this map, adjoining cells differ only in respect of one attribute. this enables us to link them up very easily into pairs at a time, four cells at a time and so on.

Thus the end cells differ from the opposite cells in respect of only 1 attribute; and the Map permits what is referred to as END-OVER adjacencies.

If you want to reduce a complicated Boolean expression into the simplest terms possible, this is a very handy device -and of course a device of this kind is just what the Engineers are looking for.

It might be an oversimplification but it could be near enough to the work, to say that Carroll's Logic, slightly improved and under another name, has recently come into its own.

Let us now try to discuss the nature of Carroll's contribution to Logic. It was as natural for Carroll to think logically as for the sparks to fly upwards. From childhood he was an inveterate devizer of games and puzzles. When most himself and least inhibited, in his stories to the Liddell children, Logic and Semantics came easily enough to provide jokes and incidents: Logic is a main ingredient in the *Alice* books. At the next stage in his life, out of concern for the administration of Christ Church, he worked out the logical theory of the committee. Events at Christ Church also led him to develop an altogether more difficult branch of Politics, the theory of parliamentary representation. As Curator of the Common Room he put Logic to tasks that common sense might have coped with better, and saved himself and others much fruitless labour. The books on Logic which he published fairly late in life were admirably adapted to instructing young people: but this was hardly a task for a logician of Carroll's calibre; and the development of the subject has by-passed his books. The logician will go to them for recreation and amusements, or to collect examples, not for guidance in logical theory.

Carroll's contribution to Logic, we safely may conclude, lies either in his indirect contribution through the *Alice* books, or in his contribution through the logical formation of political theories. Now however much fun and amusement the *Alice* books may afford logicians, and however much they may help to brighten the textbooks, it is difficult to believe that they have exerted much influence on the development of the subject. I know of no evidence that they have. If so, this leaves as Carroll's contribution, his formulation, in logical terms, of political theories.

'(Carroll's) mind, Professor Braithwaite says, 'was permeated by an admirable logic which he was unable to bring to full consciousness and explicit criticism'; and he mentions that Venn, a contemporary of Carroll's whom no one would claim to have been one of the great logicians, was 'a far better conscious logician than Carroll'.[49] Whether in Mathematics or Logic, Carroll lacked dexterity: he did not have the ability to churn out theorems, developing one from another. He also lacked the architectonic sense which would have enabled him to build up a large structure in the right proportions. He needed to keep in touch with his material and

---

[49] R B Braithwaite, 'Lewis Carroll as Logician', *Mathematical Gazette*, Vol. xvi, No. 219, July 1932, p. 176.

to feel a concern for it. And to this extent, if his contribution should turn out to be his political theories, an application of logic, arising from his experience at Christ Church, it would be in keeping with the character of the man.

Nor, if his contribution should be in applying Logic to political problems, should we be too quick to underrate this achievement: he would have been the forerunner of a recent school of Politics, whose work was separated from his by a clear seventy or eighty years - and this at a time when Logic and Politics were being studied and thought [about] in the universities on an increasing scale.

When the new school did arise, its theories were got by employing the recent methods of Economics and Logic to the needs of Political Science. Its tools had been carefully prepared over a long period, not least those furnished by the Italian economist [Vilfredo] Pareto, and were ready to be put to new uses. Carroll had got to construct his theories from first principles and without such adventitious aids.

Already, at the present day, the theory of the committee had been taken up into the new Theory of Politics as a key notion. Carroll's second bout of theorizing provided a more mature, and a highly sophisticated and difficult line of theorizing: it dealt with the choice problem of how to find the assembly that will best represent the electorate: and Carroll got much nearer to the solution of the problem than any other. When this work of his becomes accessible in the new atmosphere of thought its technique will be employed in a variety of problems and its effects will be pervasive. It is only now that Carroll may be on the point of making his contribution to the new developing Politics.

This new school of Political Science of which I speak began through economists noticing that the methods employed in solving their problems could be used also in solving political problems: the less developed Science of Politics had a sudden accession of new methods placed at its disposal. It took over the mathematical methods of Economics and at the same time began to employ formal Logic, a method of reasoning which had never been used in Economics. Already the new Politics has, in turn, begun to exert an influence on its sister subject, mainly on the branch known as Welfare Economics.

The achievements of the new science, whatever they turn out to be, lie in the future and cannot be foreseen: obviously, a great deal will depend on the abilities of its practitioners. Yet two things seem to tell in its favour. Modern science, as everyone knows, found its beginnings at a time when a group of Platonists, of whom Gallileo was the best known, applied the method of exact thought, Mathematics, and careful observation to the physical world: from this emerged Modern Physics. Nor have achievements of this kind been confined to dealing with the inanimate realm, for Economics, a human science, also entered on a new phase of development when, in the nineteenth century, it took up the explicit use of mathematical methods.

The other feature that bodes well for the new science is that though in the nature of the case their names are few, for the subject itself is barely a decade old, those whom the new Politics has attracted are, by and large, people who by their earlier work, have placed their abilities and their judgement beyond all doubt. Their names have a lustre. [These paragraphs probably date to the 1960s].

Over this new subject Carroll would reign as one of the presiding deities, sharing the honour with the Marquis de Condorcet, a French savant of the Enlightenment, who also anticipated much of the later development. Carroll might not altogether welcome the proposal and his reply might run 'I am glad to accept whatever good meanings may be attributed to my work',[50] but certainly on his day off he would want to slip down to the sea-shore to play with children, or wander round the country waiting for the last line of a poem to come into his mind.

I would suggest, however, that the invitation might be couched in terms more agreeable to him, and with a greater chance of acceptance, by pointing out that in accepting the position there might be yet another field, and one of immediate human and personal interest, over which he would be able to exercise some away. Here I can only indicate what I have in mind.

At each stage in these lectures I have given some attention to the human side of Carroll, to his hopes, desires and fears. I have also stressed that sometimes he would write, without quite knowing what subject he was writing about: he said so explicitly of *The Hunting of the Snark*. At this troubled period of his life he also wrote about committees.

Now it is a well-known feature of a scientific theory that it may apply to more than one set of phenomena. A theory developed to give an account of the flow of heat may apply equally to the flow of electricity; a theory designed of static electricity on a set of conductors, and so on.

The same, it seems to me, is true of Carroll's theories, at any rate of the theory of the committee that he developed at the same time as he was writing *The Hunting of the Snark.* It appeared to refer to committees and did so: but it referred at the same time to a subject that was still closer to him, his own feelings and thoughts. He was symbolising, it seems to me, and putting into logical form, something which cannot be expressed in the ordinary extensional Logic, but requires another form of Logic, intensional Logic, or, as we may call it, the Logic of Intensity. And Carroll's theory of the committee, I would suggest, provides this Logic of Intensity which is not to be found in the textbooks.

This is a view which I have held for some time and it may seem fanciful: but before I explain it let me give a scrap of evidence which tell in its favour. Among a batch of his papers there are some pages of rough notes Carroll made to help him to decide as between various plans for remodelling his sisters' house. And in this matter of a purely personal choice, he seems to have employed, in an elaborate

---

50 Cf. his letter to a correspondent about *The Hunting of the Snark* quoted by Stuart Dodgson Collingwood, *The Life and Letters of Lewis Carroll*, p. 173. Cf. also Roger Lancelyn Green (editor) in *Diaries*, vol. ii, pp. 352-2.

way, the technique of his theory of the committee. He used the theory of the committee to find the rational response to his own desires.[51]

To explain briefly my view that, in his theory of the committee, Carroll provided also the beginnings of a theory of intensive Logic, or the Logic of intensive quantities, let us suppose we are dealing with a committee consisting of 7 members, and that the proposals A, B, C and D are before this committee. The members' schedules are taken to be as in the diagram Fig. 1.10:

2 [or *x*]	3 [or *y*]	2 [or *z*]
A	B	C
B	D	D
C	C	B
D	A	A

Figure 1.10

2 of them prefer A to B, B to C and C to D, and similarly for the other members.

We can deduce from these schedules which motion it is rational for the committee to select. But if we are dealing with the problem of choice of the individual, exactly the same representation, or the same symbolism will apply.

Instead of taking A, B, C, D to be motions before the committee, let us take it that A, B, C, D are states of affairs which a particular individual envisages as possible, and from among which he is making a choice. His choice will be determined by his desires. Let us take it that - the individual is actuated by say, the desires *x* and *y*. The desire *x* may be directed more intensely to the state of affairs A than to B; the desire *y* may be more intense for B than for C, and so on. If so the same representation will apply in the problem of choice as in the problem of the committee: all we need do is to interpret it differently. Instead of schedules of preference for different motions, we will now have schedules of desire for various envisaged states of affairs. Instead of measuring in the vertical direction, order of preference, as in the committee problem, we now measure in the vertical direction, intensity of desire.

---

[51] [See] Warren Weaver, 'The Mathematical Manuscripts of Lewis Carroll, *The Princeton University Library Chronicle*, Vol. xvi, No.1, Autumn 1954. This does not permit any definite pronouncement, but it is possible that Carroll was trying to assign relative weights to his different desires. The desires are defined by the objects to which they refer, e.g. desire for 'better shape for stairs', desire 'not to injure view from garden', etc. Other people have made implicit use of the theory of the committee to arrive at a decision, without having constructed such a theory. An occasion on which Alfred Marshall did so is narrated by Mary Paley Marshall, *What I Remember* (Cambridge, Cambridge University Press, 1947), p. 23.

In place of the number of voters who hold the different schedules, we will have the relative weights of the different desires. We may call these relative weights *wx*, *wy*, *wz*, and we will show each relative weight at the head of the appropriate schedule.

I would not for a moment claim that this mode of representation 'solves' the problem of individual choice. But it does seem to me to display the nature of the problem in a way that makes it more intelligible than any other scheme. Difficulties remain in plenty: but when the problem has been reduced to this form, and we can see the *logical structure* of choice-making, we are at any rate nearer to the solution.

Another problem which may be tackled by the Carrollian method, is that of the probable judgement, or partial belief. This is a problem I touched upon above by way of quotations from *Wonderland*. Let us consider the dialogue between Alice and the Pigeon when Alice's neck has grown very long.

> 'But I'm not a serpent, I tell you!' said Alice. 'I'm a - I'm a -.'
> 'Well! What are you?' said the Pigeon. 'I can see you're trying to invent something!'
> 'I - I'm a little girl,' said Alice, rather doubtfully, as she remembered the number of changes she had gone through, that day.
> 'A likely story indeed!' said the Pigeon, in a tone of the deepest contempt. 'I've seen a good many little girls in my time, but never one with such a neck as that! No, no! You're a serpent; and there's no use denying it. I suppose you'll be telling me next that you have never tasted an egg!'
> 'I have tasted eggs, certainly,' said Alice, who was a very truthful child; 'but little girls eat eggs quite as much as serpents do, you know.'
> 'I don't believe it,' said the Pigeon; 'but if they do, then they're a kind of serpent: that's all I can say.'[52]

Again we are dealing with intensities, and the logical structure of the problem seems to be the same as that of the committee.

In this instance A, B, C, D stand for various propositions entertained by the mind of the individual, and he is trying to arrive at a judgement as to which of these propositions is the more likely to be true. One aspect of the situation, which we indicate by x, inclines him to the belief that the proposition A is more likely than C and C than D. Another aspect of the situation, y, seems to give greater likelihood to B being true, and so on. The facts can be depicted by the Carroll type of diagram, when we take the schedules now to refer to 'intensity of belief'.

The same representation applies in this problem as in the case of the committee; and, if so, it will be interesting to see what advance can be made in the Psychology of Probability, or the Psychology of partial belief, using the methods which Carroll hit upon almost ninety years ago.

---

52 From Wonderland, reprinted in *Complete Works*, p. 55.

[Carroll] was described by his friend Bishop Strong - one of the few friends he had at the end of his life - as 'an eccentric genius'. part of the form that his eccentricity took was that he lived to only a small extent in the world of outside observable events. One consequence of this was that when his *Diaries* were published in 1953, the reviewers sent up a whole chorus of delay. 'Was there ever such a dull dog as Lewis Carroll?' they asked. 'This man hasn't lived at all.'

They were right up to a point, because his real life was lived in a world of inner meanings. Philosophic and logical principles were just as real for him as human beings and occupied his mind just as much. the jokes in his childrens' stories often hinge on logical points or semantic difficulties. There is a preoccupation with Philosophy and Logic: he is aware of philosophic difficulties and is concerned to state them in their sharpest forms; and a logical contradiction bothers him as much as a toothache.

It may seem odd that one who was so preoccupied with the inner world, should also, if I am right, make the most distinguished contributions to the theory of political institutions. Yet the explanation is not far to seek. We know from Physics that an explanation that is worded out in connection with one branch, may be an equally good explanation in what appears to be an entirely different branch. A theory of heat may also be an equally good theory of Electricity. A theory of electricity may be equally good as a theory of Light. It just happens that the problems are isomorphic. They have the same structural properties. The equations that apply in the one case just happen to apply in the other.

This, I would suggest, is what happened with Carroll. His mind went questing for solutions to his inner problems. Where he found solutions they were solutions to his inner problems - but at the same time they were solutions to the outer and objective problems of government. His theories of government are none the worse for this. Only it means that if we are trying to find out how Carroll came to arrive at his theories, the real clues to what happened may not be the state of government in England at the time, or the state of government at his college of Christ Church: our best clues may be Carroll's state of mind at the time he was writing. He produced his political theories, it seems to me, when in the one case the political problem at Christ Church happened to coincide with the psychological problem he was confronted with, and in the other case when the political problem in England happened to coincide with a different psychological problem he had come up against.

In taking leave of Carroll, therefore, I don't feel that we are bidding good-bye to a figure going down the road, the White Knight with his horse and saddle lettered with gadgets. Rather I see him as the Beaver in *The Hunting of the Snark*, busily weaving lace in the bow. His fine-spun patterns of thought would be ready to catch up new phenomena. To some extent these would be political, as, for a short time, he had hoped. But again they would be the phenomena of mind, or spirit, and thought. They would have arisen out of Carroll's life and a certain delicacy that was his.

Part 2. The Principles of Parliamentary Representation

[The three sections of this part are respectively based on Black's papers, 'Lewis Carroll and the Theory of Games', *American Economic Review, Proceedings* 1969; 'The central argument in Lewis Carroll's *The Principles of Parliamentary Representation*', *Papers in Non-Market Decision Making* 3, 1967; and 'Lewis Carroll and the Cambridge Mathematical School of PR: Arthur Cohen and Edith Denman', *Public Choice* 7, 1970. Black's chapter plan indicates that he wanted each of them to go into this part of the book. We have put them in what seems to be the most logical order. The first section of this chapter gives an overview of Carroll's argument; the second examines it in more detail; the third gives the historical and psychological context.]

2.1. An Outline of Carroll's Argument.

Lewis Carroll's booklet, *The Principles of Parliamentary Representation*, applies the technique of the two-person zero-sum game, which we usually associate with economic theory, to provide a theory of proportional representation and a theory of the apportionment of parliamentary seats. The entire reasoning of the booklet is expressed in quantitative terms, again on the basis of the two-person zero-sum game and the maximin criterion. A few months after the booklet had been completed Carroll made a further application of game theory, this time using the coalition game, in a very tentative way, to find the most suitable set of candidates to represent a constituency in an election.

The year 1884 was the year of the last great debate in Britain on the franchise. If we exclude Ireland, the electors [almost entirely supported] one or other of the two political parties. The number of parliamentary seats was fixed, and the more seats the one party got, the fewer went to the other. This set-up invites an application of the two-person zero-sum game; but to the present day, with a solitary exception,[1] Carroll has been the only writer to provide a model of this kind.

---

[1] Jack Sawyer and Duncan MacRae, Jr., 'Game Theory and Cumulative Voting in Illinois' *American Political Science Review*, 56, 1962, pp. 937-40.

Whether we say that he actually uses the two-person zero sum game in which each of the parties acts on the maximin criterion, or on the other hand provides only a model which has the same effect as this, is a semantic question and perhaps not very important.

Carroll regards the choice of an electoral system as being a problem in proportional representation. Consider, he says, the main family of electoral systems: that in which a constituency has m seats ($m \geq 1$), and the elector is allowed v votes ($1 \leq v \leq m$), of which he can give no more than one to any candidate. The seats will of course be awarded to the m candidates with the highest number of votes. The best or most suitable electoral system will be that member of the family which gives rise to the greatest degree of representation. Or, in his treatment, going on to quantify the problem, the best electoral system will be that member of the family which maximizes 'the number of electors represented', and which consequently minimizes 'the number of electors unrepresented'. This requires, among other things, a definition of 'the number of electors represented' (or unrepresented) in any particular election; but we will come to that later.

To get his model of the two-party system in politics, Carroll must, of course, make certain definite assumptions. He supposes that each of the two political parties knows the number of its own supporters and the number of supporters of the other party; and that each party is able to direct the voting of its supporters as it chooses, getting so many of its supporters to vote for these candidates, so many for these others, and so on, with a view to maximizing the number of seats it fills. To illustrate, let us choose as a particular example a constituency contested by two parties, A with 100 supporters and B with 70; and let us take it that this constituency has three seats ($m = 3$) and the elector is allowed two votes ($v = 2$). The party A may choose to put up two candidates giving each 100 votes, i.e., may use the strategy (100, 100); or it may put up three candidates and use the strategy (100, 60, 40), or the strategy (80, 70, 50), and so on.

To tackle this problem of the three-seat two-vote constituency, we might, at the present day, argue in this fashion. The set of strategies open to the party A is the set of ways of partitioning 200 votes into two parts each consisting of 100, together with the set of ways of partitioning 200 votes into three parts no one of which exceeds 100. And conceptually we may arrange this rather long series of strategies down the left-hand column of the payoff matrix. Likewise along the top row of the matrix we may arrange all the strategies open to the party B.

Now with a matrix framed in this way, suppose that A uses any given strategy open to it and B uses any given strategy. The electoral rules will then specify which of the candidates are to get the seats, and we can fill in the figure in the corresponding cell of the matrix, showing A's payoff and B's payoff. When we fill in each cell in this way, we can proceed in the usual manner by adding a column of figures to the right of the matrix to show the minimum number of seats A fills, whichever strategy it may use, and correspondingly for B. From this, if we suppose that each of the two parties acts on the maximin criterion, we can deduce the choices made by both A and B. Approaching the problem in this way we might be apprehensive that choice by each party on the maximin criterion might not give rise to an equilibrium point -and in this problem the concept of a 'mixed strategy' would be meaningless. It can be shown, however, that in the electoral problem we need have no cause for apprehension and that, trivial exceptions apart, choice made on the maximin criterion must give rise to equilibrium.

This shows the logical structure of the problem and it is the way in which we might approach it today; but in Carroll's day a different approach appears to have been quite common. In 1884 almost two-thirds of the parliamentary seats were in multi-seat constituencies returning two, three, or four members, in which voting tactics were vitally important and came under supervision of the local party caucus. Already in 1853 in a pamphlet eulogised by John Stuart Mill (*Majorities and Minorities: Their Relative Rights)*, James Garth Marshall had shown how, in what we now refer to as the m-seat v-vote constituency, each of the two parties may choose a rational strategy. Marshall had worked out a large number of arithmetical examples and, although he did not formulate these rules, it was made fairly plain that to choose a rational (maximin) strategy it was sufficient that a party should:

(1) aim to fill a definite number of seats, s seats say ($v \leq s \leq m$).

(2) put up exactly s candidates; and

(3) distribute its votes among its s candidates as evenly as possible.

This provided a very direct way of choosing a rational strategy. Knowing its own strength and that of its opponent, a party needed to consider the outcome only if it put up v or $(v + 1)$ or ... or m candidates, dividing its votes among them as evenly as possible, while its opponent, behaving rationally, did the same from its side. Take, for instance, the three-seat two-vote constituency with 100 voters of whom x support the party A and $(100 - x)$ support the party B. A rational strategy for A would be either (x, x) or $(\frac{2x}{3}, \frac{2x}{3}, \frac{2x}{3})$ - though other rational strategies may exist which have the same outcome as one of these; and a rational strategy for B would be either $(100 - x, 100 - x)$ or $(200 - \frac{2x}{3}, 200 - \frac{2x}{3}, 200 - \frac{2x}{3})$.

Suppose x is known and equal to 46. A rational strategy for A would be (46, 46) or (31, 31, 30), 'or' being used in the exclusive sense, and for B (54, 54) or (36, 36, 36). If A puts up two candidates with 46 votes apiece, it is bound to fill one seat against the best counter strategy that B can bring against it, though the strategy (44, 42) say, would also secure this. Similarly a rational strategy for B to use is (54,

54), ensuring that it fills two seats against whichever strategy A may employ; and no strategy exists ensuring that B will fill more than two seats.

Marshall's own work was cast in terms of critical values - the minimum number of supporters required by a party to fill one or two or ... or m seats, in some arithmetical example. From the above it is easy to verify that, in the three-seat two-vote constituency, to fill one seat against the best counter strategy, a party requires the support of 40+% of the electorate, to ensure filling two seats the support of 50+% and to ensure filling three seats the support of 60+% In terms of these critical values, a party that has the support of 46% of the electorate will again be able to ensure filling one seat but not more.

Marshall's reasoning implies acceptance of what we know today as the maximin criterion in the two-person zero-sum game, and the first step in Carroll's argument is to generalize this type of reasoning for the m-seat v-vote constituency. Carroll shows algebraically the volume of support sufficient to ensure that a party which acts on the maximin criterion will fill one or two or ... or m seats, and throughout the entire chain of reasoning of the booklet the implicit assumption is that both parties act on the maximin criterion.

The next step in the argument derives from game theory and, as we would formulate it today, makes explicit use of the maximin criterion. The purpose of the booklet is to show which constituency, i.e., which values of m and v, give the maximum degree of representation. Applying Occam's razor, Carroll avoids definition of 'representation,' a complex concept which he does not require, and defines only 'the number of electors represented' or equally its converse, 'the number of electors unrepresented', for if, out of the e electors in a constituency f are represented, $(e - f)$ will be unrepresented.

His definition of the percentage of electors unrepresented is this. Suppose that in any given election in which both parties act on the maximin criterion, a party with the support of h% of the electors fills a given number of seats, but could have filled this same number of seats with the support of only h'% of the electorate $(h' \leq h)$: then the votes of $(h - h')$% of the electorate who support this party are useless or 'wasted' and play no part in determining the result, and $(h-h')$% of the electorate is unrepresented. Likewise, if the other party has the support of k% of the electors and could have filled the same number of seats with the support of only k'% $(k' \leq k)$, then a further $(k - k')$% of the electors are unrepresented. In all $(h - h')+(k - k')$% of the electorate is unrepresented.

To illustrate, take again the three-seat two-vote constituency in which one party has the support of say, 46% of the voters and, with both parties acting on the maximin criterion, fills one seat. But it could have filled one seat with the support of only 40+% of the voters. Hence 6% of the voters are unrepresented. Similarly the other party with the support of 54% of the voters fills two seats and could have done so with the support of only 50+%, so that a further 4% of the voters are unrepresented. In all in the election 10% of the voters are unrepresented.

From these two premises, the argument proceeds to establish the mathematical expectation of the percentage of the electorate unrepresented in constituencies with various numbers of seats and with the elector allowed more or fewer votes. It is found that in the *m*-seat *v*-vote constituency the mathematical expectation of the percentage of the electorate unrepresented diminishes as *m* increases and again is smaller the smaller the size of *v*. The common-sense conclusion to be drawn is that we want an electoral system based on large constituencies, with four or five seats, Carroll suggests, in which the elector is allowed only a single vote.

The model for the two-party system is got on the basis of various simplifications, one of them being the assumption that the elector prefers any candidate of his own party to any candidate of the other, and another that he is equally well represented whichever candidate or candidates his party chooses to put up. But before he wrote the booklet, in letters to the *St. James's Gazette,* and after he had got it completed, in a *Supplement* and *Postscript to Supplement,* Carroll employed a more general model than this. Suppose we have a multi-member constituency in which, in relation to the candidates who stand, the voter's preferences are subject to no restriction, and possibly do not even follow party lines. How, with *m* seats available, can we choose that set of *m* candidates which will give rise to the fullest representation?

Carroll gives what we may term (1) a conceptual approach, (2) an operational approach, and (3) a practical scheme of election. In all three he employs the Droop quota: in this *m*-member constituency it is possible (in all relevant cases) for each of exactly *m* candidates to get a full Droop quota of votes.

1. In the conceptual approach we may imagine a group of voters, each with a definite preference schedule in relation to all the candidates who stand, and each voter knowing the preference schedule of every other. We would allow the electors to form and reform themselves into coalitions, each coalition aiming to command one or two or three, etc., Droop quotas of votes, in order to fill accordingly one or two or three, etc., seats. If such coalitions were able to form and re-form by a process of contract and recontract, until a stable set of coalitions emerged from which no elector had any incentive to detach himself, we would regard the candidates returned by this set of coalitions as an optimum set of representatives for the constituency.

This would be possible in practice, however, only if (and of course here and elsewhere we are using modern terms where they help to express Carroll's notions) the costs of obtaining information about the preference schedules of other voters, and about the state of the coalitions at any moment, were zero or very low, and if the costs of entering into fresh contracts with the members of any coalition were zero or were very low, and if the process could be carried out within a short period of time. In fact, except for the very small group, the costs of the elector finding out the preferences of other electors, finding out the existing state of the rival coalitions, and of terminating the existing contract and entering into a new contract with another coalition would be far from zero. It is just not feasible in the ordinary election for the voters to form and re-form themselves into coalitions, one set after another, until a stable set is arrived at which will return *m* candidates.

2. We may therefore seek an operational approach. Take it that we have a given set of preference schedules. The voters cannot in practice enter into coalitions among themselves, but we may be able to draw up rules whereby we can discern the coalitions which the voters, acting rationally, might form. This would show which individual candidates would get the support of one Droop-quota coalitions to elect these individual candidates, which pairs of candidates would get the support of two Droop-quota coalitions to elect these pairs of candidates, and so on.

Carroll in fact sketched a few rules which would assist in picking out coalitions of this sort, but said they were incomplete; and after working longer on the problem he probably got to know that, in general, the answer to the problem is indeterminate. But, in particular examples, determination of the coalition which it would be appropriate for the voters to form seems clear enough to common sense, and Carroll gives two or three instances of this sort: in fact he employs them to show that the single transferable vote gives a different and therefore a wrong answer in these particular cases. In general, however, an operational answer to the problem is again not feasible.

3. In the practical scheme of election which he proposes for use in Great Britain, Carroll side-steps both the previous difficulties. He argues that, in the first place, the average British elector, 'Hodge, fresh from the plough!' will know which candidate he prefers to any of the others, but will be unable to rank the candidates in order of preference: his felt preferences will amount to him knowing only which of the candidates he likes most. If so, the elector should be asked to give a vote only a single candidate. Then at the end the election the candidates themselves, Carroll suggests, should collect up the votes and treat these votes as if they were their own private property. Any candidate with a Droop quota of votes should elected. When this had been done the next stage would be for the candidates to meet and exchange their votes among themselves, those who had already be elected exchanging or donating their surplus votes over and above the Droop quota which had been used in their election. It would be at this stage in the process that the coalitions would be formed. The candidates would already know another's political attitudes and, with only a few well-informed and well-practised people meeting together after an election, the transaction costs in the formation of coalitions would be very low. Thus in casting only first-preference ballots the voters would have expressed accurately their attitudes; and, at the same time, the choice of an optimum set of candidates as envisaged by the conceptual approach, would have been attained by the actual formation of coalitions - but coalitions among the candidates and not among the voters; i.e., among people whom coalition formation is appropriate.

We will not attempt to evaluate this suggestion of Carroll's. For our present purposes we wish only to point out that just as in the two-party contest he makes use of the two-person zero-sum game, so in the non-party or multi-party case Carroll makes use of the basic notions of what we know today as the coalition game with ordinal utilities.

2.2. The Central Argument in *The Principles of Parliamentary Representation*

Introductory. The preface to the first edition of Carroll's *The Principles of Parliamentary Representation* is dated Guy Fawkes' day, 1884, and that of the second New Year's Day, 1885. The suggestion may have been to abolish the old and establish the new, but since then only a single author[2] has shown any understanding of the booklet. This is the more to be regretted if, as in the opinion of the present writer, the booklet is the most interesting contribution to Political Science that has ever been made, and is one which may yet prove fertile of future developments. We will confine our attention to the central argument of the booklet, Carroll's theory of PR, omitting his theory of the apportionment of seats to districts of different sizes, and the scheme of election which he thought suitable for adoption by Great Britain at the time.

Before giving an account of Carroll's theory we would allude to two features in the historical setting. Apart from the group of Irish Nationalists sitting in the house, members of parliament at the time belonged either to the Liberal or the Conservative party; and, on the face of it, any discussion of British electoral arrangements was entitled to assume the existence of a two-party system. The other feature was that in the general election of 1880 70% of the MPs were returned by multi-member constituencies, the bulk of them by two-member constituencies; and out of the total of 658 members, 36 were returned by the so-called 'three-cornered constituencies' in which the elector felt aggrieved at being given only two votes instead of the three needed to let him have one vote for each seat to be filled. (Incidentally we examine this type of electoral system in the arithmetical example given below). Also in the course of the article we will allude briefly to some of the authors to whose work Carroll appears to have been under obligation.

It is usually easier to follow an argument if we know the kind of result to which it leads up. Carroll's main purpose, in his theory of PR, is to find the mathematical expectation of the percentage of the voters represented in various electoral systems, and his results are summarized in the table[3] of Figure 2.1. He will show, for example, that if a parliamentary constituency has five seats and each elector is allowed four votes, each vote to be given to a different candidate, on the average of an infinite number of elections 61% of the voters will be represented by the five members who get the seats: or if, in this five-member constituency, the

---

2 Walter Ernest Smith, *Fair Representation, an Essay* (London, Kegan Paul, Trench & Co., 1885, pp. 63). Partly through his knowledge of Carroll's theory, Smith was himself able to provide an interesting discussion of PR, but his only references to Carroll's work are unfavourable (*op. cit.* pp. 31 and 39--40).

3 The earlier version of the table in the article 'Redistribution', *The St. James's Gazette*, 11 October 1884, extends the figures as far as the ten-member constituency in which the elector may have one or two or ... or ten votes.

elector is allowed say, only one vote, then on the average of an infinite number of elections 84% of the voters will be represented by the five members chosen. No other theory of representation has been so far-reaching in its scope as this.

Mathematical expectation of the percentage of the electorate represented

	Number of votes each elector can give					
Number of members assigned to each constituency	1	2	3	4	5	6
1	51					
2	68	51				
3	76	66	51			
4	81	74	64	51		
5	84	79	71	61	51	
6	86	81	74	66	59	51

Figure 2.1

The assumptions. To get a rigorous theory Carroll must, of course, make definite assumptions, and the assumptions he in fact makes are as follows:

(i) There are only two political parties.

(ii) In any constituency each party knows the number of its own supporters and the number of supporters of the rival party.

(iii) Each party is able to get its supporters to cast their votes in whatever way the party prescribes, some of its supporters voting say, for the candidate a_1, others for a_2, and so on, or perhaps some of them voting for a_1 and a_2, while others vote for a_2 and a_3, etc.

(iv) An election to fill, say, three seats will choose the candidates who have the highest, the second-highest and the third-highest numbers of votes; and in general those candidates are elected who have the highest numbers of votes, until the available seats have been filled.

(v) the electoral system may be such that a constituency has one or two or three or ... seats; and if it has say, five seats, the elector may be allowed five votes to be given to different candidates, or (as another distinct possibility) four votes to be given to different candidates, or ... or only one vote. Thus Carroll's theory will cover all the cases in which the constituency has m members ($m \geq 1$) and the elector is allowed v votes ($1 \leq v \leq m$), but is restricted to giving only one out of his v votes to any particular candidate.

Carroll's theory may have owed the wide sweep of electoral systems that it covers to what appears to have been the first work in PR (apart from articles in magazines and newspapers) that he read, the scholarly pamphlet by W R Ware, *The Machinery of Politics and Proportional Representation*[4] which reprinted an article that had appeared a little earlier in *The American Law Review*. Ware's work furnished[5] a classification of electoral systems that showed the various cases which might arise, and Carroll was able to bring all of the main electoral systems under the one 'general' theory.

The electoral contest viewed as a two-person zero-sum game with the players acting on the maximin criterion. In Carroll's model one party with full knowledge of the voting possibilities confronts another with like knowledge. The more seats the one party gets, the fewer will be filled by the other and the interests of the two parties are strictly opposed.

The payoff to each party, the number of seats it fills, will depend, as well as on the number of its supporters, on the strategy it adopts, i.e. on the number of candidates it puts up and on the instructions it gives to its supporters about voting for them. Its payoff will also depend on the strategy adopted by its rival. Carroll's assumption here is that each party takes it that whatever strategy it adopts, its rival will use the most effective counter-strategy: he envisages each of the two parties as acting on what we know today as the *maximin criterion*. His model for the election is the two-person zero-sum game.

One of the *Alice* books is based on a game of cards, the other on a game of chess: but whereas the *Alice* books may be equally enjoyable whether or not we pay any attention to the underlying game, his booklet on elections can be understood only in terms of the strict logic of game theory.

A particular example chosen to illustrate the three steps of the argument. Carroll's purpose is to provide the entries for the cells of the table of Figure 2.1. If we know how to get the figure for one cell, any other can be filled in using the same technique, and it will be sufficient to show how to fill in one cell in the table.

To do this we choose any particular electoral system, say that in which the constituency has three seats and each elector is allowed two votes to be given to different candidates. We will take it that the total number of electors in the constituency is 100. It is required to compute the number of electors represented, on the average of an infinite number of elections, in this electoral system, i.e. in the three-seat two-vote constituency. To make the arithmetic slightly simpler we will in

---

4 W R Ware, *The Machinery of Politics and Proportional Representation* (London, Representative Reform Association, 1872, pp. 31)

5 *Op. cit.* pp. 8--15.

fact compute the number of electors who are *unrepresented*: then if the number of electors unrepresented is say, 46, the number represented will be 100 - 46 = 54.

The computation for this or any other cell in the table consists of three distinct steps:

(a) finding the number of seats that a party fills, with any given number of supporters, when it acts on the maximin criterion

(b) for any given distribution of support between the two parties, finding the total number of voters represented; and

(c) assuming a given probability distribution of support between the two parties, finding the mathematical expectation of the number of voters represented.

As a preliminary to step (a) we require two lemmas.

Lemma 1. If a party aims to fill a precise number of seats, an optimum strategy for it is to put up just that number of candidates.

Suppose the party A aims to fill precisely one seat. On the assumptions we make the party can get its supporters to vote as it directs, and of course we disregard contingencies such as illness, misreading the ballot paper, etc.. If the party A cannot fill one seat when it puts up one candidate, neither can it do so if it puts up two or three candidates. Likewise if the party aims at filling precisely two or precisely three seats, an optimum strategy is for it to put up the corresponding number of candidates. Or if it aims to fill zero seats, as effective a way as any is for it to put up zero candidates.

Lemma 2. An optimum strategy for a party which aims to fill any given number of seats is for it to distribute the available votes as evenly as possible among that number of candidates.

Suppose that in the three-seat two-vote constituency we consider, the party A has say, 31 supporters: it has 62 votes which it can distribute, giving not more than 31 to any candidate. Let this party aim to fill precisely three seats. And suppose that if it gives its candidates a_1, a_2, a_3, say 25, 21 and 15 votes receptively, it will in fact fill three seats. The candidate a_3 is elected with 15 votes.

We can safely assume that, putting up a given number of candidates, the party will not transfer votes away from any candidate without transferring them to one of its other candidates. If in the instance we have chosen it transfers the votes among its candidates so as to make the distribution more even, then each candidate will still receive at least 15 votes and all three will be elected. If the three candidates would be elected on a less even distribution of votes, they will necessarily be elected on a more even distribution.

Obviously a more even distribution of votes, say (24, 20, 16) or (23, 20, 17) or ... or (21, 21, 20) would, in certain circumstances enable the party to fill precisely three seats, whereas a less even distribution would not do so. And in general an optimum strategy for filling any precise number of seats is for the party to put up just that number of candidates (Lemma 1) and divide its votes as evenly as possible among them: if it cannot do so in that way, no other distribution of votes would enable the party to fill the precise number of seats at which it aims.

More exact proofs could be given of Lemmas 1 and 2, but these indications are enough for our present purposes.

(a) The minimum number of supporters needed to fill one or two or three seats in the constituency. Recurring to the example of the three-seat two-vote constituency, it is required to find the minimum number of supporters the party A must have to fill one seat or two seats or three seats, when the opposing party uses the best strategy which is open to it to restrict the number of seats that A fills. It is also required to show that if a has the minimum number of supporters needed for it to fill say, two seats, but not three seats, acting on the maximin criterion A will in fact fill two seats.

To fill one seat. Let A have x voters at its disposal, and let us consider how small x can be and still be sufficient to enable A to fill one seat when B takes the best counter-action open to it. Since each elector has two votes to be given to different candidates, if, with x supporters, A puts up either one candidate or two candidates, it can give to each x votes.

The party B has $(100 - x)$ supporters. Its most efficient policy to prevent A from filling a seat will be to put up three candidates, just enough to fill the available three seats, and instruct some of its supporters to vote for its candidates b_1 and b_2 say, others to vote for b_2 and b_3 and others to vote for b_1 and b_3, in such a way that, so far as possible, each of its candidates gets the same number of votes apiece.

The party A will be just able to fill one seat if it is just able to prevent the other party from filling all three seats; that is, if the number of votes for a candidate of A is just greater than the number of votes per candidate for each of three candidates of B;

$$\text{that is, if } x \text{ is just greater than } \frac{2(100-x)}{3}$$

that is, if x is just greater than 40.

Hence, assuming the other party to adopt the best strategy open to it, to fill one seat the party A must have at least 41 supporters.

To fill two seats. If A aims to fill two seats, an optimum strategy for it to do so is to put up two candidates and instruct each of its supporters to give one of his two votes to each candidate. An optimum counter-strategy for B, in attempting to prevent A from filling two sets, will also be to put up two candidates and instruct each of its supporters to give one vote to each.

With these instructions to its x supporters, the number of votes per A-candidate will be x. With these instructions to its $(100 - x)$ supporters, the number of votes per B-candidate will be $(100 - x)$.

A will be just able to fill two seats if it is just able to prevent the other party from filling two seats; that is, if the number of votes per candidate of A is just greater than the number of votes per candidate of B:

that is, if x is just greater than $100 - x$

that is, if x is just greater than 50.

Hence, to be just able to fill two seats, A must have at least 51 supporters.

If the party A aims to fill all three seats, an optimum strategy for it is to put up three candidates and distribute the $2x$ votes at its disposal as evenly as possible among them.

The best counter-strategy for B will be either to put up one candidate, instructing each of its $(100 - x)$ supporters to give one of their votes to him, or to put up two candidates, instructing its supporters to give one vote to each candidate: in each case the number of votes per B-candidate is $(100 - x)$.

The party A will be just able to fill three seats if it is just able to prevent the other party from filling one seat; that is, if each of its three candidates gets just more votes than a candidate of the other party;

that is, if $\frac{2x}{3}$ is just greater than $(100 - x)$

that is, if x is just greater than 60.

Hence, assuming the other party to use the best counter-strategy, in order to be able to fill three seats the party A must have the support of at least 61 of the voters. We could have deduced this from a previous answer. If one party must have just over 40 supporters to capture one seat, the other, to capture all three seats, must have just over 60.

All cases can be dealt with arithmetically along the above lines. For example, if we take a constituency with five seats in which the voter is allowed three votes to be given to different candidates, we can compute the minimum percentage of support that a party must have in order to fill one, two, three, four, or five seats. Alternatively we could get the answer in any such case by substitution for the variables in the algebraic inequalities given by Carroll (See below, p. 159).

Lastly we prove that if, as in the Carroll model, both parties act on the maximin criterion, a party with any given number of supporters will fill precisely the number of seats shown by the above type of arithmetic, and that, for instance, in the two-party three-seat two-vote constituency, if A has say, 56 supporters, it will fill two seats. In these circumstances A acting on the maximin criterion, takes it that whatever strategy it may use, B will employ a counter-strategy which gives to it, A, the least favourable result. Now B, with 44 supporters, has open to it a strategy whereby it can fill one seat, irrespective of the strategy that A may employ. Hence acting on the maximin principle, A will not aim at filling more than two seats but will aim to fill two seats. And, from the arithmetic, it controls a sufficient number of votes to enable it to do so. In general, a party acting on the maximin criterion will fill the number of seats given by the above type of arithmetic.

The figures will of course sometimes fall on the dividing line between two different cases and then the result will be indeterminate. This would happen in the constituency we consider if one party had exactly 40 supporters and the other exactly 60, or if both parties had exactly 50 supporters apiece. Here it is sufficient to point out this possibility without attempting to investigate it further. In a parliamentary constituency with thousands of voters a coincidence of this kind would have a very tiny probability of occurrence, and strictly speaking it is to the large constituency that the figures apply which Carroll goes on to compute.

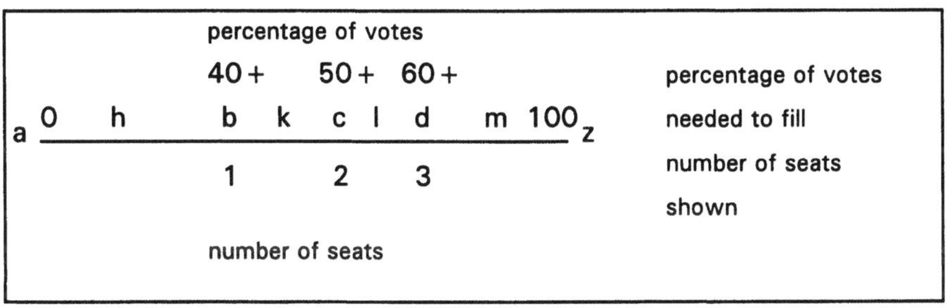

Figure 2.2

Our findings for the two-party three-seat two-vote constituency are summarized in figure 2.2. The party will fill zero seats if it has the support of less than the fraction $\frac{ab}{az}$ of the electorate: it will fill one seat with a support between $\frac{ab}{az}$ and $\frac{ac}{az}$: it will fill two seats with a support between $\frac{ac}{az}$ and $\frac{ad}{az}$ it will fill three seats with a support between $\frac{ad}{az}$ and $\frac{az}{az}$. It will be found that a limiting case in which a party's support coincides with one of these fractions, e.g. is exactly $\frac{ac}{az}$, does not influence the figures that we go on to compute.

The type of argumentation used in the above arithmetic and used by Carroll in arriving at the algebraic inequalities to which we have referred was first employed by James Garth Marshall in Appendix C of his pamphlet *Minorities and Majorities: their Relative Rights: a Letter to the Lord John Russell, M.P., on Parliamentary*

Reform (See below, pp. 175--79), and there is a fair presumption that Carroll would be acquainted with Marshall's pamphlet.

(b) The Baily-Carroll definition of 'the number of voters unrepresented'. We will take it, in the geometrical argument that follows, that in the three-seat two-vote constituency which we consider the electorate runs into thousands: this avoids the need to deal with discrete quantities. For this or any other case, the definition that Carroll employs of 'the number of voters unrepresented' is one of striking simplicity. Suppose that the party A has *ah* supporters and B has *hz* supporters (Figure 2.2). The *ah* supporters of A fail to fill a single seat: they are unrepresented. The *az* supporters of B fill all three seats. But, says Carroll, they could have filled these three seats with only *bz* supporters and *therefore hb* of the supporters of B are unrepresented. Since this is true at whatever point in *ab* the point of division h between the strengths of the two parties lies, we can conclude that provided $a<h<b$, the total number of electors unrepresented is *ab* and the fraction unrepresented is $\frac{ab}{az}$.

Next suppose that the point of division between the strengths of the two parties lies at k in Figure 2, $b < k < c$. A will fill one seat: but it could have filled one seat with only *ab* supporters, so that by the same argument as before *bk* of A's supporters are unrepresented. B will fill two seats: it could have filled them with only *cz* supporters; and *kc* supporters of B are unrepresented. Hence for $b < k < c$, the total number of electors unrepresented is *bk* supporters of A *plus kc* supporters of B, in all *bc* electors and the fraction unrepresented is $\frac{bc}{az}$.

By the same argument if the point of division between the strengths of the two parties lies at *l*, $c<l<d$, *cl* supporters of A and *ld* supporters of B, in all *cd* electors are unrepresented, the fraction of the electorate unrepresented being $\frac{cd}{az}$. Or, if the cut between the parties lies at *m*, $d<m<z$, the number of electors unrepresented is *dz* and the fraction unrepresented is $\frac{dz}{az}$.

To formulate a definition corresponding to this usage, suppose that in a particular election in Carroll's model a party with the support of x voters fills a given number of seats, when the minimum number of voters needed to fill these seats is y, $y \leq x$; then y voters of this party are represented and $(x - y)$ supporters of the party are unrepresented.

The reader may well feel this definition of 'the number of electors unrepresented' to be unsatisfactory; but to attempt a satisfactory definition of the term in the present place would blur the outlines of our account of Carroll's theory. We will accept here Carroll's usage of the term, and it should be borne in mind that it is this definition that underlies the percentages shown in Figure 2.1.

Carroll had taken over the definition from the pamphlet by Walter Baily, *Proportional Representation in Large Constituencies* (Ridgway, London, 1872, pp. 22). This was a direct borrowing.

(c) Computing the mathematical expectation of the percentage of the voters represented: the argument in probability. From the overall point of view the number of voters who happen to be represented or unrepresented in a particular election is of limited significance. The important question is: Given any electoral system, what proportion of the voters is represented, or unrepresented, on the average? and this means on the average of an infinite number of elections: or, in other words, What is the *mathematical expectation* (M.E.) of the fraction represented or unrepresented?

Fraction of the electorate unrepresented	Probability
$\frac{ab}{az}$	$\frac{ab}{az}$
$\frac{bc}{az}$	$\frac{bc}{az}$
$\frac{cd}{az}$	$\frac{cd}{az}$
$\frac{dz}{az}$	$\frac{dz}{az}$
Total	1

Figure 2.3

We have established that in our three-seat two-vote constituency the fraction of the voters unrepresented will be $\frac{ab}{az}$ or $\frac{bc}{az}$ or $\frac{cd}{az}$ or $\frac{dz}{az}$, according as the cut between the two parties lies in *ab* or *bc* or *cd* or *dz* respectively. We set out these fractions in the left-hand column of Figure 2.3.

Next Carroll assumes that the point of division between the strengths of the two parties is *equally likely to fall at any point in az*, i.e. assumes that (in the large constituency with which we are dealing) the division of strengths between the two parties is given by a rectangular probability distribution. If so, the probability that the cut between the strengths of the two parties will lie in any particular segment of *az* will be proportional to the length of that segment. The probability that the cut lies in $ab = \frac{ab}{az}$, that it lies in $bc = \frac{bc}{az}$, that it lies in $cd = \frac{cd}{az}$, that it lies in $dz = \frac{dz}{az}$. These probabilities give the right-hand column in Figure 2.3.

From Figure 2.3 the M.E. of the fraction of the electorate unrepresented is

$$\frac{ab^2}{az^2}+\frac{bc^2}{az^2}+\frac{cd^2}{az^2}+\frac{dz^2}{az^2}=\frac{ab^2+bc^2+cd^2+dz^2}{az^2}$$

When we disregard small quantities in the calculation this gives for the three-seat two-vote constituency,

M.E. of the fraction of the electorate unrepresented

$$= \frac{40^2 + 10^2 + 10^2 + 40^2}{100^2} = 34 = 34\%$$

Hence the M.E. of the fraction of the electorate represented is 66%.

Fraction of the electorate represented	Probability
$\frac{bz}{az}$	$\frac{ab}{az}$
$\frac{ab+cz}{az}$	$\frac{bc}{az}$
$\frac{ac+dz}{az}$	$\frac{cd}{az}$
$\frac{ad}{az}$	$\frac{dz}{az}$
Total	1

Figure 2.4

Alternatively, from Figures 2.2 and 2.4,

M.E. of the fraction of the electorate represented

$$= \frac{ab.bz + bc(ab+cz) + cd(ac+dz) + dz.ad}{az^2}$$

which, after reduction,

$$= 1 - \frac{ab^2 + bc^2 + cd^2 + dz^2}{az^2}$$

The cells in the table of Figure 2.1 can all be filled in the manner we now have fully demonstrated, but we give two further illustrations.

Example. The single-seat single-vote constituency.

Whichever party has the greater number of supporters fills the seat. The minimum percentage of votes needed to fill the seat is just over 50%, i.e. 50+%. If for instance one party has the support of 67% of the electorate, on the Baily-Carroll definition 50+% of its supporters are represented and (67 - 50+) = 17-% of its supporters are unrepresented. The 33% of the electorate who support the other party are unrepresented, and it is easy to see that in this or any other election a total of 50-% of the electorate will be unrepresented and 50+% represented. Since this is true in every single election, the M.E. of the percentage of the electorate represented (in this instance irrespective of the probability distribution assumed) will be 50+%, which Carroll shows in the table as 51% (Figure 2.1).

Example. The five-seat two-vote constituency.

percentage of votes							
0	29	34	50	67	72	100	percentage of votes needed to fill
	1	2	3	4	5		number of seats shown
number of seats							

Figure 2.5

We can turn to Carroll's table (Table III, see p. 160) showing the minimum percentage of support needed to fill one, two, ..., five seats and this gives the percentages shown in our figure 2.5(a); or we can compute the percentages either arithmetically or using Carroll's algebraic inequalities and this gives the percentages shown in figure 2.5(b). Taking the figures of figure 2.5(b),

$$(100)^2 \text{ M.E. of fraction of electorate unrepresented}$$

$$= \left(28\frac{4}{7}\right)^2 + \left(33\frac{1}{3} - 28\frac{4}{7}\right)^2 + \left(50 - 33\frac{1}{3}\right)^2 + \left(66\frac{2}{3} - 50\right)^2 + \left(71\frac{3}{7} - 66\frac{2}{3}\right)^2 + \left(100 - 71\frac{3}{7}\right)^2$$

$$= 2\left[\left(\frac{200}{7}\right)^2 + \left(\frac{100}{21}\right)^2 + \left(\frac{50}{3}\right)^2\right]$$

$$= 2234$$

and M.E. = 22%, this figure being again slightly more accurate than the 21% that would be shown as unrepresented by Figure 2.1 above.

The probability distribution assumed. It might be objected that, contrary to Carroll's assumption, not all party strengths lying between 0 and 100% of the electorate are equally likely to occur: the assumption of a rectangular distribution runs counter to our experience of a two-party system and to our expectations arising from a theory of political parties. Thus for the electoral system corresponding to Figure 2.2 or that corresponding to Figure 2.5, few people would regard the cut between the parties as being equally likely to occur at any point in *az*, but would regard it as being more likely to occur, for either of the electoral systems concerned, nearer the centre of az rather than towards the extremities. The general view would be that while the probability distribution of party strengths would vary from one electoral system to another, it would tend to be bell-shaped and not rectangular.

This is in no sense an objection in principle to Carroll's theory. it requires only that we determine as well as we can, both from the wider theory of a party system and from such empirical evidence as is available, the probability distribution that we think appropriate to whichever electoral system we are dealing with, and amend the calculation accordingly.

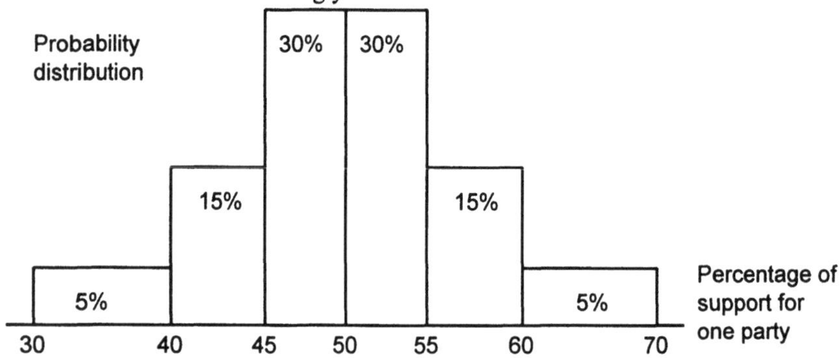

Figure 2.6

Percentage of the electorate unrepresented	Probability
40	.05
10	.45
10	.45
40	.05
Total	1

Figure 2.7

Suppose, for instance, that for an electoral system based on the three-seat two-vote constituency the estimate arrived at was that the probability distribution of party strengths would be as in Figure 2.6. The probability computation would then be that displayed in Figure 2.7, giving

M.E. of the percentage of the electorate unrepresented

$$=(40 * .05 + 10 * .45 + 10 * .45 + 40 * .05) = 13\%$$

The probability distribution of party strengths would not be independent of the type of electoral system in use but would be a function of it, being different for say, the electoral system based on the single-seat single-vote constituency and that based on the three-seat two-vote constituency. Even the number of political parties would depend on the electoral system in use. But obviously no writer is expected to solve all problems at once, and these are not the problems that Carroll is attempting to solve.

Implications of the table of Figure 2.1 for the choice of an electoral system. Now that we have shown how to arrive at them, let us accept the figures of the table of Figure 2.1 - which implies of course accepting the assumptions and the Baily-Carroll definition of 'the number of voters represented' on which the table rests. The *desideratum* (and we accept this for the moment as the *only* objective to be aimed at) is to maximize the mathematical expectation of the number of electors represented. Looking along any row in the table we notice that for a constituency with two or more seats the smaller the number of votes allowed to the elector, the greater is the percentage of the electorate represented, and the percentage represented is a maximum when the elector is allowed a single vote. Whatever the number of seats in the constituency, therefore, the elector should be allowed only a single vote.

Looking down the first column in the table, i.e. taking it that each elector is in fact given a single vote, we observe that the greater the number of seats in the constituency, the larger is the percentage of electors represented. 'The obvious conclusion is - let the Districts be as *large* as possible, and let each Elector give *one* vote only'. But, Carroll points out (and the percentage of electors who are unrepresented is immediately seen from figure 2.1), 'the change from single-Member to two-Member Districts changes the percentage of unrepresented Electors from 49 to 32 ...; whereas the change from five-Member to six-Member Districts, only changes the percentage from 16 to 14 .... The conclusion is that *the* important point is to have as few single Member, and even as few two-Member, Districts as possible; but that, when we have got as far as to Districts returning four or five Members each, it is hardly worth while to go further'. Keeping in mind practical considerations (which he does not mention) such as the difficulties of the electors getting to know the candidates in a large constituency of their programmes, and the difficulties of the representatives keeping in touch with the electors, Carroll opts for an electoral system with four- or five- seat constituencies, in which the elector is allowed a single vote.

2.3. Carroll and the Cambridge Mathematical School of PR: Arthur Cohen and Edith Denman

[*The Principles of Parliamentary Representation*] provides a quantitative theory of proportional representation. It does so by interpreting the two-party system in politics in terms of Game Theory; and, since Game Theory was not officially invented until the appearance of a paper by John von Neumann in 1928, the booklet's lack of success is not to be wondered at. The only branch of Game Theory involved in the booklet itself is the two-person zero-sum game, and, with the texts now available, it is not difficult to get a knowledge of this, the simplest as well as the most fundamental branch of Game Theory, and the maximin criterion on which it is based. But in Carroll's day to follow an argument in terms of the two-person zero-sum game would have needed considerable mathematical ability; the main concepts had not been isolated and named; they were difficult to identify and elusive to use, and moved about in the problem anonymously. To add to the difficulty Carroll sets out the argument in a very summary manner, in a way to satisfy himself rather than the reader, and omits to give simple but necessary explanations without which the argument is almost unintelligible.

In four letters, written to *The St. James's Gazette* in May and June 1884 before the appearance of the booklet, and again in the *Supplement* to the booklet and the *Postscript to Supplement* which appeared in quick succession in February 1885, Carroll employs a different games concept, that of the coalition game, but he does so only at the intuitive level and breaks off in his argument rather abruptly. On the other hand *The Principles of Parliamentary Representation* makes consistent use, from first to last, of the two-person zero-sum game.

A possible genesis of the booklet. We [have shown in Part I] that Carroll's earlier work on the theory of committees had been a response to various circumstances at Christ Church - to its committees, to Dean Liddell's architectural plans for the college, and above all [that] his theory of committees had been a way of symbolizing the desires the fears of his love for Alice Liddell. It is possible that something of the same may be true of *The Principles of Parliamentary Representation.*

The most obvious aspect to follow up was the parlour games element, for we know that from childhood Carroll was an addict both of playing and devising games. In the garden at Croft he had devised his railways game with its very elaborate and firm rules. Later he had threatened resignation from Curatorship of the Common Room unless the committee would obey the rules that he imposed. But none of the games he constructed makes use of the maximin criterion. His fondness of games tells little about the booklet.

The Electoral Board (May 1882). The Christ Church side is more revealing and the minutes of the Governing Body together with Carroll's diary entries suggest what appears to be a beginning for his theorizing on proportional representation. By a revision of the Christ Church statutes,[6] an Electoral Board was to be constituted to exercise important powers in regard to the appointment of Students, their emoluments, etc. 'The appointment of ... Students shall rest with an Electoral Board, consisting of the Dean and four other persons, to be elected annually at a Stated Meeting to be held on a day fixed by the Governing Body' (Statutes, section XVI).

---

6 Cf. Sir Charles Edward Mallett, *A History of the University of Oxford* vol. iii (London, Methuen, 1927), pp. 330--53, Henry L Thompson, *Christ Church* (London, F E Robinson, 1898) p. 207 and Thompson, *Henry George Liddell*, pp. 146--7.

The Dean explained the points that would need action and the governing body reached its decision:

> Any one having an absolute majority of the votes of those present and voting shall be declared duly elected. If after the final voting it appears that none or only a part of those voted for has obtained an absolute majority, a second vote shall be taken; and so on, till each of the 4 has obtained an absolute majority. After each voting the number of votes obtained by each person voted for shall be read out by the Chairman (minute of the Governing Body, 17 May 1882).

Carroll at this time commits much of his thought to his diary.[7]

> *My 17 (W). 1882. The question came on in the Governing Body Meeting, as to a mode of taking votes for the 'Electoral Board' of 4, which, by the new Ordinance, we have to elect. I proposed a scheme, devised while lying awake last night, the principal object of which was to enable a minority to get *one* member in. It was, in essence, to let every one give *4* votes, all to one candidate or separately, and consider any one elected who gets more than 1/5 the total number of votes. My scheme was not even seconded. The one proposed by J B Thompson was adopted - 'Each write 4 names, and any one getting an absolute majority of votes to be elected' - i.e. any one getting 13 votes from 25 electors - which may happen with 7 candidates. My belief is that the rejected method would be more just, in the interest of *minorities*.
>
> *May 31 (W). 1882. Governing Body Meeting. Election of Electoral Board. I suggested two difficulties that might arise, but they would not consider them {and he works out arithmetical examples to illustrate the difficulty}.
>
> Neither of my difficulties occurred in voting.
>
> *June 10 (Sat). 1882. Special Meeting of Governing Body .... As to mode of taking votes on recommendations of Electoral Board next Thursday, J B Thompson opposed Ince's proposal of balloting (which I seconded), and spoke exactly on the opposite side to the line he took on May 17. Ince's proposal was carried.

---

7 Entries kindly provided by the late Miss F Menella Dodgson, for which it had not been possible to find room in the published *Diaries*, are shown by *. The other entries are to be found in Green, *The Diaries of Lewis Carroll*, vol. ii, at the appropriate dates.

Some years earlier Carroll's proposals about the procedure for taking votes on the design of the belfry had been circulated to members beforehand and his suggestion had been adopted. Now the Governing Body was confronted with the abstruse issue of proportional representation and , without benefit of visual aids, discussion must have been difficult. Carroll and Thompson, the Lee's Reader in Anatomy, were distinctly critical of each other;[8] and, with a logical point at issue and a strong case against Thompson, Carroll had no inclination to relent.

Lawn Tennis Tournaments (August 1883). After a gap of a year he wrote four letters on tennis tournaments to *The St. James's Gazette* and reprinted one of them as a pamphlet.[9] The mode of thinking is the same as in the theory of committees (1873--76), the two problems being isomorphic.

A challenge. Carroll had attempted to 'write up' his theory of committees in a pamphlet he had printed in 1876, but regarded the outcome as unsatisfactory; yet he still believed that, rummaging at the roots of things, he would be able to find a better solution than any he had so far obtained. With a view to further examination of the subject, in December 1877 he had sent to a number of colleagues in the other colleges interleaved copies of the pamphlet along with 'the cyclostyled sheet' asking for:

> an account of any rules, written or unwritten, adopted in your College to settle difficulties arising in elections ...
>
> A really scientific method for arriving at the result which is, on the whole, most satisfactory to a body of electors, seems to be still a *desideratum*.

The franchise debate came both as a challenge and an opportunity - a challenge to bring to fruition a line of thinking which, so far as he knew, he alone had cultivated, and an opportunity to find an audience on a theme that had some fascination for him. The political dispute aroused old feelings and this was propitious to effort; however abstract his thought might be, it had always to arise out of a substratum of feeling and emotion; it was this that made him a poor mathematician and a great thinker.

The Franchise Bill 1884, and Redistribution of Seats 1885. By the beginning of 1884 the discussion of electoral reform had become nation-wide. The Great Reform

---

8 For evidence of earlier disagreement and their quarrel some months later, see Hudson, *Lewis Carroll*, pp. 245--7.

9 Cf. Roger Lancelyn Green, 'Lewis Carroll and The St. James's Gazette', *Notes and Queries*, 7 April 1945, and Green (ed.), *The Lewis Carroll Handbook*, pp. 110--11. The pamphlet is reprinted in Alexander Woollcott (ed.), *The Complete Works of Lewis Carroll*, pp. 1201--11 and in Roger Lancelyn Green (ed.) *The works of Lewis Carroll* (London, Paul Hamlyn, 1965), pp. 1059--66.

Bill of 1832 had admitted the merchanting and industrial classes to the franchise, formerly confined to the landowners. Disraeli's Act of 1867 gave the vote to the workers in the towns, and early in 1884 Gladstone introduced a Bill to give the vote to agricultural labourers and miners and he promised to redistribute parliamentary seats in the following year.

The proposed franchise created wide differences of opinion even within the parties. Yet the most glaring deficiency in representation was not in the franchise itself but in the maldistribution of parliamentary seats. For instance, at one end of the scale a group of boroughs with an aggregate population of .25 million returned 42 members, and, at the other end, practically the same number, 43 members, were returned by boroughs with an aggregate population of 6.73 million. The Act of 1867 had made little attempt at the equalization of constituencies, and since that time the disparities had grown wider through increase of population in the north while the south remained static[10].

Gladstone's Bill passed the Commons to be rejected in the Lords. The deadlock was broken later in the year when the leaders of the two parties met and agreed on a scheme of redistribution. The Franchise Bill became law in December, and a Redistribution Act in June of the following year chopped up the country into districts of [very] roughly equal population. The multi-member constituencies, which hitherto had predominated, were largely swept away in favour of the single-member constituency.

So completely did the multi-member and multi-vote constituency disappear from England that it may be well to remind ourselves that the circumstances in which Carroll was writing were very different from those of later times, and that a wide range of electoral procedures were then in use. We may do this briefly by the table[11] of Fig. 2.8 which shows that of the 658 seats in the parliament of 1880, no fewer than 422 (almost two-thirds of the total) were in multi-member constituencies returning two, three, or, in the case of the City of London, four members. In the three-member constituency the elector was allowed two votes which he could give either to a single candidate or to two candidates, as he chose.

---

10 Cf. H J Hanham, *Elections and Party Management, Politics in the Time of Disraeli and Gladstone* (London, Longmans, 1959), pp. x--xi and 403--4. Cf. also Charles Seymour, *Electoral Reform in England and Wales, The Development and Operation of the Parliamentary Franchise, 1832--1885* (New Haven, Yale, 1915), pp. 348--9.

11 The figures have been compiled from *Dod's Parliamentary Companion*, 49th year (London, Whittaker, 1881), pp. 116--57, 54th year (London, Whittaker, 1886), pp. x--xi and 403--4. Cf., also Seymour, *Electoral Reform in England and Wales*, pp. 348--9)

Distribution of Multi-member and Single-member Constituencies in the General Elections 1880, 1885 and 1886[12]

		Number of constituencies returning				
General Election	Number of Constituencies	1 member each	2 members each	3 members each	4 members each	Number of members (seats)
1880*	422	196	211	12	1	658
1885	643	616	27	-	-	670
1886	Identical with 1885					

Figure 2.8

The public would be further familiar with the multi-member constituency from the mode of election to School Boards which was in operation 1870--1902: 'At every such election every elector shall be entitled to a number of votes equal to the number of members of the School Board to be elected, and may give all such votes to one candidate, or may distribute them among the candidates as he thinks fit' (Elementary Education Act, 1870, section 29). In 1884 the discussion about the method of election appropriate to School Boards continued unabated and much of it was well-informed. Their experience of electoral conditions was richer and wider than has been available since, and the writers and public of Carroll's day would be in a better position to appreciate the variety of electoral conditions discussed in the booklet.

Foundation of the PR Society at the beginning of 1884. In January 1884 a group of influential publicists and lawyers joined forces to form the Proportional Representation Society which was formally constituted in March. They wisely decided, right at the start, to put aside whatever minor differences of opinion they might have as to which scheme of PR was *precisely* the best, recognising that if several schemes were put forward the effect would be to confuse the public and give

12 In the general election of 1880 the number of members actually taking seats was 652, instead of 658 as shown in the table, due to the disfranchisement of the two-member constituencies, Bridgwater and Beverley, and the single-member constituencies, Cashel and the borough (as distinct from the county) of Sligo, Cf. Sir Thomas Erskine May, *Treatise on the Law, Privileges, Proceedings and Usage of Parliament*, 9th ed. (1883), p. 26, *The Parliamentary Poll Book*, 7th ed. (1910), Part I, pp. 334--6 and 34-, J F S Ross, *Parliamentary Representation*, 2nd ed. (1948), p. 289, 3rd ed. (1955), p. 466, H J Hanham, *Elections and Party Management*, p. 398 and Cornelius O'Leary, *The Elimination of Corrupt Practices in British Elections 1868--1911* (Oxford, Oxford University Press, 1962), p. 235.

it a distaste for the whole notion of PR, and they agreed on unanimous support of the single transferable vote [s.t.v.] in the form which makes use of the Droop quota. Soon they were conducting a campaign in the newspapers and magazines and a number of eminent speakers gave public lectures and some leaders of the society conducted trial elections to demonstrate the operation of the s.t.v.. Their efforts were particularly successful among members of parliament and by April the society had enrolled over 180 members belonging to both parties.[13] It looked as if, in Great Britain, some form of PR might become a reality.

Carroll's four letters to The St. James's Gazette, 15 May--5 June 1884. In May Carroll began a sustained attempt to arrive at a satisfactory scheme of PR. To all appearances his first two letters to the *Gazette* were merely an attempt to discredit the scheme of the PR Society by discovering examples in which the s.t.v. would select the wrong candidates. But his third letter showed that behind the simple facade there was much more than this, indeed that his attempt was to find the basis for a general theory of PR. The fourth letter was again an arithmetical example to illustrate a case in which he considered the s.t.v. would choose the wrong candidates. The nature of his efforts can be seen from the diary entries.

> *May 2 (F). 1884. Spent yesterday afternoon with Baynes {Rev. Robert Edward Baynes, who lectured on Mathematics, afterwards Lee's Reader in Physics at Christ Church} in calculations on subject of Proportionate Representation.

> *May 15 (Th). 1884. My letter on 'Proportionate Representation' appeared in the *St. James's Gazette*. An article by Sir J. Lubbock [the founder of the PR Society] was in the *Daily News* this morning, in which he talks of the chance of the wrong man coming in on their system as 'microscopical' and 'infinitesimal'. In my instance it exceeds $^1/_2$!

> *May 16 (F). 1884. an answer to my letter, by Mr. Arthur Cohen, appeared in the *St James's*.

Cohen's letter is neat but it presupposes that the principle lying behind the s.t.v. is right, which was exactly the thing that Carroll wanted to question.

On May 17 a letter from another correspondent, W C Sidgwick, appeared in the *Gazette*. Sidgwick complains quite rightly that Carroll is using a model which is inappropriate in the problem of PR. Later we will suggest a reason for Carroll's use of a wrong model at this stage.

---

13 Cf. *First Annual Report of the Proportional Representation Society* (1885) and Clarence G Hoag and George H Hallett, Jr., *Proportional Representation* (New York, Macmillan, 1926), pp. 164--7 and 179--81.

> '{Mr. Dodgson} demonstrates,' writes Sidgwick, 'that this method {of the single transferable vote} will cause candidates to be elected who are not desired by the majority of the electors. He does not, however, seem to have observed that this is precisely the object with which the scheme is put forward.'

> *May 17 (Sat). 1884. Wrote a reply to Mr. Cohen's letter on 'PR'.

His reply to Cohen, for reasons which will become clear later, was studiously polite. At the same time Carroll replied more brusquely to Sidgwick, disclaiming the mistake that he had attributed; but his third letter went on to sketch a wider theory of PR which would be free from the fallacy that Sidgwick had noticed.

May 23--27, Carroll spent the weekend in London.

May 27, a third letter appeared, written before his departure for London.

5 June, a fourth letter appeared in the *Gazette*, which we mention a little out of the temporal sequence; it continues 'the charge which I brought ... against the method of voting proposed by the 'Society for Proportionate Representation,' that it is liable to bring in the wrong man'. It attempts, rather ineffectively, to sum up the argument of his three earlier letters by providing an example in which the s.t.v. chooses the wrong set of candidates and this example is incorporated in *The Principles of Parliamentary Representation.*

After the return from London a new model in the article of 5 July. As soon as he got back from his weekend in London Carroll sought a new approach, but quickly changed his mind about it when he discovered something better.

> *May 29 (Th). 1884. A new plan for 'Proportionate Representation' occurred to me yesterday, and I have now worked it into a very hopeful shape. It is to have *one* Member only for each constituency, and to give him voting-power in the House, as follows:- if he has '*a*' supporters, and '*b*' vote for next man (or, if he is sole candidate, against him), his power = (2a - b)/1000 taking integer nearest to this, unless it be $^1/_2$, when power must be '1'.

The situation Carroll was confronting was that some boroughs returned a member for a population of 5,000 while others returned one member for 150,000 of population. He was also trying to take into account the greater or smaller majorities that the members got.

> *June 2 (M). 1884. I have come to the conclusion that varied voting-power will not do. It would require (if we gave all Liberals, equal power, and ditto for all Conservatives) that each man's power should vary *inversely* as the number of members in his party, which would never be endured. And it would not mend matters to divide the total voting-power of the party on

other proportions among its members. The best plan seems to be to give each district one member for every *m* electors in it, and to use the Society plan, modified as I have suggested. (*m* = number of electors in kingdom divided by number of members in House).

*June 3 (Tu.). 1884. Concocted a new 'Proportional Representation' scheme, far the best I have yet devised, and sent the MS to Mr. Greenwood, to add to a letter now in print. The chief novelty in it is the giving to each candidate the power of transferring to any other candidate the votes given for him.

*June 4 (W). 1884. Wrote a new version of above scheme, which I hope Mr. Greenwood will print as an article: it is too long for a letter. Fourth letter appeared. {In fact it appeared on 5 June.}

*June 9 (M). 1884. He *has* printed it as an article, but wants me to abridge it.

*June 10 (Tu.). 1884. Sent a condensed version of the article to *The St. James's Gazette.*

*July 5 (Sat). 1884. Appearance in *The St. James's Gazette*, of my article on 'Parliamentary Elections'.

*July 10 (Th). 1884. Mr. A. Cohen approves article.

Up to this stage the diaries are a great help in following Carroll's views on PR. In May--June 1882 he had made three entries, two of them setting out schemes of PR. In 1884 for 2 May--4 June there are eight entries, some indicating possible schemes of PR; the five entries for 9 June--10 July refer to the writing or printing of his article and to Mr. A. Cohen's approval of it. But after this there is no further mention of PR until 23 September.

Carroll's work before end-May 1884 and again in Feb. 1885 was a natural development from his earlier theory of committees, uninfluenced by the magazine literature. While he was Curator, Carroll took a detailed interest in the magazines to which the Common Room subscribed and went to the length of arranging to bind together articles from earlier issues which he considered might be of interest to other members. The magazines taken by the Common Room,[14] including *The Pall Mall Gazette*, *The Spectator*, *The Athenaeum*, *The Contemporary Review*, *The Fortnightly Review*, *The National Review* and *The Nineteenth Century*,

---

14 Information got from the minutes which Carroll wrote as Curator (Christ Church Common Room).

in 1883--84 contained a variety of articles on PR, a subject of general interest at Christ Church. The main feature which emerges from an examination of this literature, however, is of a negative kind. The magazine articles would provide some stimulus to take up the problem again and express his own quite different views on the matter; but during May 1884 and again at the time he was writing the *Supplement* and *Postscript to Supplement*, neither the magazine articles nor any other literature (if we exclude that on the Droop quota) appears to have had any influence worth mentioning on Carroll's thinking. At these times the line of thought he pursues is to take a given group of preference schedules (just as in the theory of committees) and examine, in imagination, the coalitions which, under various hypothetical conditions, would be likely to form. The search is for consilience, or rank correlation, among the preference schedules of the voters, when these schedules are arranged as different sets of given sizes. This work along the lines of the coalition game is the natural extension of the theory of committees which he had obtained some years earlier. It is the most fundamental way of looking at the whole problem, with the drawback only that it quickly gets too complex to yield useful results.

The switch from the coalition game to the two-person zero-sum game early in June 1884. Taking into account the diary entries, we would judge that the three letters which he published during May represent fairly fully Carroll's thinking at the time, and he tried to sum up this line of thought in the arithmetical example of his letter of 5 June, which makes use of the coalition game. On 2 June he saw the practical scheme of election that was the logical outcome of his coalition model.

At this point there is a sudden switch in Carroll's way of looking at PR and the two-party system. On 4 June he did a first version of the article 'Parliamentary Elections' which was to appear a month later, sketching some of the main positions of the booklet. The coalition game has vanished, to be replaced, as the model in terms of which he is thinking, by the two-person zero-sum game. There are other changes in this article which are only less significant. Carroll now seeks the 'principles' or 'axioms', to be termed in the booklet the *desiderata*, which, to be satisfactory, an electoral system should be able to satisfy. Indeed the article is arranged as a sequence of five such axioms, three of them of a formal nature, and one of them 'That the waste of votes, caused by accidentally giving one candidate more than he needs, and leaving another of the same party with less than he needs, should be if possible avoided' is given a key role in the booklet.

The switch from the coalition game to the two-person zero-sum game early in June 1884. Taking into account the diary entries, we would judge that the three letters which he published during May represent fairly fully Carroll's thinking at the time, and he tried to sum up this line of thought in the arithmetical example of his letter of 5 June, which makes use of the coalition game. On 2 June he saw the practical scheme of election that was the logical outcome of his coalition model.

At this point there is a sudden switch in Carroll's way of looking at PR and the two-party system. On 4 June he did a first version of the article 'Parliamentary

Elections' which was to appear a month later, sketching some of the main positions of the booklet. The coalition game has vanished, to be replaced, as the model in terms of which he is thinking, by the two-person zero-sum game. There are other changes in this article which are only less significant. Carroll now seeks the 'principles' or 'axioms', to be termed in the booklet the *desiderata*, which, to be satisfactory, an electoral system should be able to satisfy. Indeed the article is arranged as a sequence of five such axioms, three of them of a formal nature, and one of them 'That the waste of votes, caused by accidentally giving one candidate more than he needs, and leaving another of the same party with less than he needs, should be if possible avoided' is given a key role in the booklet.

Now it is easy enough to trace the provenance of some of the ideas which Carroll already holds early in June and expounds in the booklet. In adopting the model of the two-person zero-sum game he has switched to the argument which had been introduced by James Garth Marshall. In the article of 5 July Carroll has a thorough grasp of Marshall's method of reasoning which was later to provide the backbone for the booklet, and it is possible that at the time of composition of the article he had already seen Marshall's pamphlet; but whether or not he done so is made difficult to judge by the fact that, both among th practising politicians of the local caucus (which at the time had reached the peak of its development) and among the political writers and commentators, Marshall's way of looking at the problem was fairly well known. Indeed Marshall's technique had been employed in an article in *The Contemporary Review*[15] for May 1884 which Carroll would have read in the Common Room. By early June he knew Marshall's technique, but whether he got it directly from Marshall's pamphlet or in some other way we cannot be sure.

We can be fairly certain, however, from a comparison of the texts of the two writer, that already in June Carroll had taken the method of *desiderata* and the concept of 'wastage of votes' direct from the pamphlet by Walter Baily, and, we may mention, Baily's work had been much influenced by H R Droop.

Why did Carroll change over from the one type of theory to the other? Carroll puts forward two distinct theories of PR, the one worked out in the booklet,[16] based on the two-person zero-sum game, and the other sketched in the four letters to the *Gazette* and in the *Supplement* and *Postscript*, based on the very general type of reasoning of the coalition game. Two conclusions that he had reached from his more general type of reasoning - his refutation of the single transferable vote and his practical scheme of elections - got incorporated in the booklet where they are presented without the logical basis that they need and are treated as if their justification were self-evident - and in a booklet which otherwise offers a strictly connected chain of reasoning, this is quite out of place.

---

15 The Representation of Minorities' by G Shaw Lefevre, *op. cit.* pp. 714--33. See Shaw Lefevre's footnote at p. 731.

16 And before that in the two articles 'Parliamentary Elections', *The St. James's Gazette*, 5 July 1884, and 'Redistribution', *ibid.*, 11 October 1884.

It is clear that we will understand the booklet much better if we can find an explanation for Carroll starting out with one type of argument, moving to another, and then switching back again to the first. The date on which he forsook the coalition argument and moved to the other can be located, from the diaries, to within a day or two; [to] about 4 June. The cleavage is perfectly sharp.

This feature was very puzzling to me until I came on some documents which seemed to show the whole of Carroll's theories in a new light. Carroll had sent out copies of his booklet to all members of the Proportional Representation Society (diary entries of 5 Nov. and 23 Dec. 1884) and I had known for some time that a copy of the booklet was in the library of the society. It seemed possible that the copy might have been sent by Carroll himself and, if so, he might have sent an accompanying note with some comment. The Vice-President of the society, Mr. John Fitzgerald, kindly conducted me through the archives for the period. There was no letter from Carroll but we came on a note from Arthur Cohen whose letter had appeared in the Gazette, and two other notes which had some bearing on the booklet. Cohen's letter[17] is addressed to Samuel Insull, organising secretary of the society.

Dec. 10 [1884]

Dear Sir,

I am very sorry that I have to attend a public meeting at 8 o'clock on Monday evening, which will prevent my attending the meeting of which you are good enough to give me notice.

I will do my best to attend the next, as I think the subject is of the greatest importance and interest.

Mr. Charles Dodgson of Christchurch Oxford, has asked me to send him a list of the members of the proportional representation Society. He has written a very learned and ingenious pamphlet on the subject, and is desirous of sending a copy to each of the members. I would be glad if you could send me a list of them, so that I could forward it to him.

Yours truly,

Arthur Cohen

In the copy of Insul's reply[18] to Cohen a few words are faint and illegible.

Dec. 17 1884

Dear Sir,

---

17 In a miscellaneous collection 'Letters 1884--1906'.

18 Letter Book No. ii, folio 158 of the PR Society.

In reply to your note our Clerk is copying a full list of our Members for Mr. Dodgson of Oxford ... a week ago .... If you require another copy kindly let me know.

Yours truly,

Samuel Insull

Carroll, getting impatient, must have written to Insull, who sent a somewhat different list to him direct

19 Dec. 1884

Enclos

Dear Sir,

I enclose you the list of the names and addresses of those Members of our Society who are not Members of Parliament as requested.

You should have had them earlier but we have been so very busy with our meetings.

Yours truly,

Samuel Insull

At this stage in our narrative we must turn aside to a quite separate chapter in the history of PR which may provide some explanation for Carroll's changeover from one type of theory to the other.

2.3.1. The Cambridge Mathematical School of PR and Arthur Cohen

The Cambridge School. Anyone who glances through the technical literature of PR in England before the end of 1884 may notice a singular feature - the large proportion of it due to writers who had studied Mathematics at Cambridge University: and, of these mathematicians, a fair proportion had proceeded to the study of law at Lincoln's Inn (although the library of Lincoln's Inn contains no publication on PR of the period before 1885). There are notable exceptions: James Garth Marshall, whose studies were at Edinburgh, Thomas Hare, of the Inner Temple, John Stuart Mill, Robert Lytton (Edward Robert Bulwer Lytton, afterwards first Earl of Lytton) educated at Harrow and Bonn, Sir John Lubbock, the first president of the Proportional Representation Society, who had forsaken Eton for a banker's desk at the age of fifteen and whose *Representation* (Swan Sonnenschein and Co., pp. 90) appeared in June 1885, and Frederic Seebohm, also a banker; but of these only Marshall and Hare made any technical contribution, and it is with the technical literature, trying to treat PR in an exact way, that we are mainly concerned.

Of the writers on PR in the following list all had studied Mathematics at Cambridge, all except Nanson had proceeded to the law, and a majority had been admitted at Lincoln's Inn. Archibald Smith had died in 1872 and Nanson was in Australia, but all of the others were active in support of the Proportional Representation Society at its commencement and four of them were members of its executive committee. Another Cambridge man, Albert Henry George Grey, afterwards the fourth Earl Grey and Governor General of Canada, who had been first in the Law and History Tripos in 1873 and a Fellow of Trinity, was Treasurer of the society and worked in close collaboration with the group. Our brief notes[19] do not attempt to show whether those who entered parliament did so before or after 1885 and we mention only main contributions to PR of a technical kind.

Archibald Smith, Glasgow University and Trinity College, Cambridge; 1836 senior wrangler and 1st Smith's prizeman; Fellow of Trinity; Lincoln's Inn; gold medallist of the Royal Society. Author of *What is the best Form of Cumulative Voting? considered in a letter to Thomas Hughes, Esq., M.P.* (Spottiswoode & Co., London, 1867, pp. 8).

John Westlake;[20] Trinity; 1850 6th wrangler and 6th classic; Fellow of Trinity; Lincoln's Inn and M.P.; Whewell Professor of International Law at Cambridge.

Arthur Cohen; Magdalene; 1853 5th wrangler; Inner Temple; chairman of the Bar Council; M.P., F.B.A. Cohen's only publication on PR was the letter to *The St. James's Gazette* that we have had occasion to mention.

Henry Richmond Droop; Trinity; 1854 3rd wrangler; 1855 1st class in the Moral Science Tripos; Fellow of Trinity; Lincoln's Inn. Author of *On Methods of Electing Representatives* (Macmillan & Co., 1868, pp. 32), 'On the Political and Social Effects of Different Methods of Electing Representatives', *Papers read before the Juridical Society*, vol. iii, part xii, March 1869, pp. 469--507 (reprint published by William Maxwell & Son, London); *Proportional Representation as Applied to the election of Local Governing Bodies* (Wildy & Sons, London, 1871, pp. 35); 'On Methods of Electing Representatives', *Journal of the Royal Statistical Society*, vol.

---

19 Information from W W Rouse Ball and J A Venn (editors), *Admissions to Trinity College, Cambridge*, 5 vol. (Cambridge, Cambridge University Press, 1911 etc.); J A Venn (ed.), *Alumni Cantabrigienses, Part II, From 1752 to 1900*, 6 vol. (Cambridge, Cambridge University Press, 1940 etc.); Joseph Foster, *Men-at-the-Bar* (London, Reeves and Turner, 1881); H W C Davis and J R H Weaver (eds.), *Dictionary of National Biography 1912--1921*, (Oxford, Oxford University Press, 1927); Frederic Boase, *Modern English Biography*, 6 vol. (Truro, 1892--1921), and the published admission registers on the Inns of Court.

20 Cf. *Memories of John Westlake* (London, Smith, Elder, 1914), chapter iv, 'Public Affairs' by Lord Courtney.

xliv, part ii, June 1881, pp. 141--202. The first of these (at page 17) introduces 'the Droop quota' and his last paper is still a good introduction to the theory of PR. He realised that PR needed a firmer philosophy than that provided by John Stuart Mill, and arranged for the translation from the French of the pamphlet by Ernest Naville, *On the Theory and Practice of Representative Elections* (Wildy & Sons, London, 1872, pp. 4 + 32) to which he contributed a preface.

To speak in a little more detail, H R Droop (1832--84) came on his father's side from a North German family. He had been third wrangler in a year in which Edward John Routh was senior wrangler and James Clerk Maxwell second wrangler and Smith's prizeman. 'In due course he was elected Fellow of Trinity, and appointed a mathematical lecturer. He voluntarily resigned his Fellowship from an honourable scruple about its retention when his private means enabled him to dispense with the emolument as a means of support. His lectureship he gave up that he might study law. He became an accomplished and scientific conveyancer, and attained as much practice as he cared for. But his favourite studies were ecclesiastical post-Reformation law and the theory of proportional representation'.[21] He exerted a wide influence on the theory of PR through his publications, his correspondence with scholars at home and abroad, and again through his influence on Walter Baily.

Leonard Henry Courtney,[22] afterwards Baron Courtney of Penwith; St. John's; 1855 2nd wrangler and Smith's prizeman; Fellow of St. John's; Lincoln's Inn and M.P.; Professor of Political Economy in the University of London 1872--75.

Henry Fawcett;[23] Peterhouse and Trinity Hall; 1856 6th wrangler; Fellow of Trinity Hall; Lincoln's Inn; Professor of Political Economy at Cambridge 1863--84, and predecessor in that chair of Alfred Marshall; M.P. and Postmaster General.

Walter Baily; St John's; 1860 2nd wrangler; Fellow of St. John's; Lincoln's Inn; government inspector of schools; Vice-President of the Physical Society of London; inventor; served on the council and closely associated in later life with University College, London. Author of *A Scheme for Proportional Representation* (Ridgway, London, 1869, pp. 12) and *Proportional Representation in Large Constituencies* (Ridgway, London, 1872, pp. 22) and the latter was to have a profound influence on Carroll's booklet.

---

21 Obituary notice, *The Times*, 22 March 1884. The writer, probably Leonard Courtney, goes on to pay suitable tribute to Droop's devotion to PR in the last year of his life. The library of the PR Society (now the Electoral Reform Society) has preserved a few letters from Droop which give a singularly clear impression of the man.

22 Cf. G P Gooch, *Life of Lord Courtney* (London, Macmillan, 1920).

23 The letter written in 1858 to Fawcett by W Hopkins, the famous Cambridge coach, is, in regard to the origins of the Cambridge School, intriguing but quite inconclusive. Cf. Leslie Stephen, *Life of Henry Fawcett* (London, Smith, Elder, 1885), pp. 48--51.

Charles Thomas Mitchell; Caius; 1865 4th wrangler; Fellow of Caius; Lincoln's Inn.

Edward John Nanson; Trinity; 1873 2nd wrangler and 2nd Smith's prizeman; Fellow of Trinity; 1875--1922 Professor of Mathematics at Melbourne, Australia. Author of 'Proportional Representation', *Transactions of the Royal Society of Victoria*, 1880, and the pamphlet *Electoral Reform, an Exposition of the Theory and Practice of Proportional Representation* (T E Verga, Melbourne, 1899, pp. 16).

James Parker Smith; Trinity; 1877 4th wrangler and 2nd Smith's prizeman; Fellow of Trinity; Lincoln's Inn; M.P., Secretary of State for the Colonies and Privy Councillor. Author of *Preferential Voting: the Transfer of Superfluous Votes* (dated 26 Feb. 1884), reprinted as appendix ii, pp. 336--42 in John H Humphreys, *Proportional Representation* (1911); Carroll refers to the pamphlet at page 30 of the booklet. Parker Smith played a large part in drafting some of the early literature of the PR Society and the clarity of the evidence he gave at a much later date before the Royal Commission on Systems of Election is likely to be remembered by anyone who reads it.[24] Among much else it gives the first formulation of 'the cube law' for a party system.

These Cambridge mathematicians-turned-lawyers, had a common educational, social and business background and were linked up in 'the network'. They had opportunity to meet and discuss their main interests: they knew one another's views on PR and were acquainted with one another's writings; and among the few letters of the time which have survived in the archives of the PR Society, are some in which they discuss and compare their views. Among some members of the group were ties of close friendship and the personal links were sometimes

[24] Cf. *Minutes of Evidence of the Royal Commission on Systems of Elections*, Cd. 5352 of 1910, pp. 77--86. The Scottish Family Smith has too much interest for Social Science to pass by without comment. James Parker Smith was the eldest son of Archibald, mentioned above, and the booklet *Fair Representation, an Essay* (Kegan Paul, Trench & Co., London, 1885, pp. 63) by the second son, the Rev. Walter Ernest Smith who had studied at New College, Oxford, has dull stretches but shows clear signs of brilliance. Furthermore W. E. Smith is the only writer who has shown any understanding of Carroll's work on PR and who has benefited from it, though this did not inhibit him from making disparaging references to it.

The family would appear to have no connection with that of Adam Smith. It originated in the Smiths of Craigend in Stirlingshire and had consistently shown high mathematical ability over a number of generations and was notable also in public life; an uncle of Archibald's had been provost of Glasgow and James Parker Smith sat in parliament for a Glasgow constituency for sixteen years. His other brothers included 'that brilliant Treasury official, ... Sir H Babington Smith, ... Mr. Arthur Hamilton Smith, the distinguished antiquary, who was for many years Keeper of Greek and Roman Antiquities at the British Museum, and ... Brigadier-General E G Smith.' (*The Times* obituary notice of James-Parker Smith, 1 May 1929).

strengthened by family connections; Westlake was the son-in-law of Thomas Hare and Droop was married to Walter Baily's sister.

Their intellectual gifts are not in doubt and almost any member of the group might be held up as an exemplar of Victorian 'character' at its best; they mustered determination, conscientiousness, tireless industry and unflagging devotion to the public good. They felt that PR, giving due rights to minorities, was an essential part of electoral justice; and, over a long span of years they showed an undefaced energy and the temper of reform. Courtney sacrificed the prospect of a seat in the cabinet for his beliefs.

But they were busy men, deeply committed in other directions, and only spasmodically did they give their time and energy to PR - except in the case of Courtney, who never felt the urge to make a theoretical contribution, and H R Droop, for both of whom PR was in the nature of a vocation. For the most part the publications of the group were made at times of general interest.

PR at their hands did not have the opportunity of academic development. the professor in his study can prepare lectures or write articles, with the encouragement of an audience and the guarantee of a circle of readers. He can expect the audience to be critical and informed about the literature of the subject. He can take some propositions to be known to his readers, without the need to explain everything between the covers of a single article. He subjects himself and his colleagues to the discipline of his subject, and expects a like discipline on the part of his students. He has the chance to think about his subject continuously and not merely spasmodically. Textbooks are written to embody the common way of looking at things; the unsolved problems can be identified, and a division of labour used in their solution. The concepts of the subject become clarified and purified, and experiment can be made in their mode of use. The Cambridge group had none of these advantages and remained one of gifted amateurs.

We can, without too much violence to the facts, add at least two names to the list and still refer the group as the Cambridge Mathematical School of PR. James Garth Marshall (1802--73) had no connection with Cambridge but he belonged to an earlier generation and his pamphlet *Minorities and Majorities: their Relative Rights* provided a beginning to the mathematical type of reasoning on PR which the Cambridge School continued and developed. Thomas Hare (1806--91) had received an honorary degree from the Cambridge and is easy to fit into the group because he exerted strong influence through his book *The Election of Representatives, Parliamentary and Municipal* (1st ed. 1859, 4th ed. 1873) and numerous other writings, through his friendship with some members of the group and through his son-in-law Westlake, and, as we have mentioned, the inner circle of the PR Society had agreed to place its full support behind Hare's single transferable vote.

In the theory of PR at the time when Carroll was writing, the Cambridge School had established a number of positions which were, however, isolated from each other and did not fit into any overall pattern. The Droop quota had been well understood. Hare's single transferable vote had been worked into the orthodox form which it was to retain. The work of James Garth Marshall (on the minimum number of voters needed to elect one, two, ... representatives in various electoral systems) had seeped into the magazine articles and become part of the stock-in-trade of the local party caucus. Owing to the unsatisfactory state of probability theory at the time, the treatment of the transfer of ballot papers from the surpluses of candidates who had already obtained a quota, was rather unenlightening. Walter Baily had a mathematical intuition of the direction that a theory of PR might take, but it remained an intuition and was far from being worked out. The school had put out some attractive buds but was still far from flowering.

Arthur Cohen (1829--1914). The member of this school whom we are most interested, Arthur Cohen,[25] was a descendant of two of the most prominent Jewish families in England. After attending De Morgan's classes in Mathematics at University College, London, he was, through the influence of the Prince Consort, the first Jew to be admitted to Cambridge, matriculating at Magdalene College. After a very active social life, which included the presidency of the Cambridge Union in his final year, Cohen in 1853 was fifth wrangler, and proceeded to the Inner Temple. He achieved an early prominence and advanced to become one of the foremost members of the English bar; he represented Great Britain in various international arbitrations, sat on royal commissions, was a member of parliament, became a Fellow of the British Academy and a privy councillor. The dedication of Dicey's *Conflict of Laws* pays the tribute that 'his mastery of legal principles was surpassed only by the kindness with which his learning and experience was placed at the service of his friends.'

A characteristic of Cohen's, it seems, was the extreme care with which he would work through the documents of a case, and the exhaustiveness of his knowledge of the literature of a branch of the law. Another was that his passion for Mathematics, beginning in childhood, continued throughout his life; we may discount the stories of the tender age at which Newton's *Principia* had been his favourite reading, but it is certain that, even when an old man, he continued to enjoy as a recreation reading works in Physics and Mathematics, making his own annotations. Writing to Mrs. Dicey after Cohen's death, Dicey says: 'As I look back at well nigh forty-four years which have passed since he made acquaintance with me on the ground of his liking my book on *Parties to Actions*, it comes back more and more upon me how much help and appreciation he gave me. It is little likely that ... I shall write any further law book. And in a sense I am almost glad that this is so, for

---

[25] Cf. J A Venn, *Op. Cit.*, *Dictionary of National Biography*, Lucy Cohen, *Arthur Cohen, a Memoir by his Daughter for his Descendants* (London, Bickers & Son, 1919, pp. vi and 216 plus two genealogical tables), and including the article 'The Right Hon. Arthur Cohen, K.C. (1830--1914)' by A V Dicey, reprinted from The Law Quarterly Review, January 1915.

half of the interest that, for instance, my *Conflict of Laws* gave me was due to Arthur Cohen's interest in it and the counsel and help he gave me in writing it'.[26]

Already when his letter appeared in the Gazette, Cohen would be known to Carroll not merely as a public figure; he had been a friend and pupil - one of the three best pupils - of 'Uncle Hassard' (Hassard Hume Dodgson, Master of the Court of Common Pleas) with whom Carroll's relationship had always been very close.[27] At an early stage in Cohen's career Uncle Hassard had again been of service to him by way of writing a strongly favourable testimonial,[28] and Cohen was not the man to forget an obligation.

Cohen may have advised Carroll on the literature. We now want to consider some of the facts which have emerged, though admittedly choosing only a selection of them and the particular selection we make will bias the conclusion.

(i) About 4 June Carroll abandoned a line of thought which had been the natural extension of his work on committees, and adopted in place of it the Cambridge approach, particularly as it is to be found in the pamphlet of 1872 by Walter Baily.

(ii) Carroll received a letter from Cohen on 10 July and on 10 December Cohen made his enquiry on Carroll's behalf to the PR Society.

(iii) Cohen refers to Carroll's booklet, which had been published a month earlier, as 'very learned and ingenious'. But on the face of it 'learned' is about the last adjective one would apply - the booklet itself gives no grounds for this - unless one knew something about the manner of its composition.

(iv) Already in May Carroll would be aware that his earlier line of thought had proved intractable, that failing some breakthrough which had not yet taken place, each further step would be attended by the greatest difficulty, and that he might be approaching a dead end.

(v) Cohen's qualities were known to Carroll through his Uncle Hassard. He was well equipped and had shown his willingness to give advice.

---

26 Robert S Rait, *Memorials of Albert Venn Dicey* (London, Macmillan, 1925), pp. 228--9.

27 'He read with the well-known pleader, Mr. Dodgson, for whom he had a great admiration. It seems to have been reciprocated' (Lucy Cohen, *Op. Cit.* p. 18). When Hassard Hume Dodgson died, in September 1884, Stuart Dodgson Collingwood was still a boy, and Collingwood's biography fails to appreciate Carroll's friendship with his uncle.

28 Lucy Cohen, *Op. Cit.* p. 47.

And if he did in fact give advice, it would be of a bibliographical nature.

Not to labour the evidence, for the matter must remain open to speculation and conjecture, it would seem to us likely that it was through Arthur Cohen that Carroll got to know the work of Walter Baily - and if of Baily, very probably also that of other members of the Cambridge School - and that this advice began before 4 June. It may have been given in correspondence between them, but it would be more in character[29] for Carroll to visit Cohen; and our own belief is that he did so on the long weekend 23--27 May that he spent in London.

The absence of acknowledgement in the booklet would not be difficult to understand, for Cohen was a Liberal member for Southwark and his views on the best form of PR differed from those of Carroll; and, in any case, in the nineteenth-century contention, acknowledgement of obligation was far less frequent than at a later date. More difficult to reconcile with our view is the comparative absence of reference to Cohen in the diaries.

But whether in this or some other way, Carroll had been brought into contact with the work of the Cambridge School. He was absorbed into a line of thought that had been developing over three decades. The Cambridge School had carried out some of the preparatory work. Carroll got some essential ingredients from the earlier writers and provided the rest himself. He used some of their work, but his system was a new creation rather than a culmination of the work of his predecessors. Where there had been only isolated propositions, he left a consistent theory; where there had been scattered fragments of groundwork, he left a completed edifice. He presented a closed system, without chink or cranny.

2.3.2. Edith Denman

The analytical evidence. On the analytical side and knowing about Carroll generally, but without any knowledge of his specific circumstances at the time, we might surmise that some, if not all, of his work on PR had reference to a division in his own nature, of the same kind that arose in regard to his attraction to Alice Liddell. He took up in May 1884 the general apparatus of his theory of committees (which had developed as a symbolism for the earlier affair) and attempted to adapt it to the problem of PR. The significant feature about this early wok, as Sidgwick, the correspondent to the *Gazette*, had pointed out, is that it does not deal at all with the problem of PR, but expresses the attitude of 'Those who wish the majority ... to return all the members'. Carroll replied (second letter, 19 May) that in the case he had discussed 'Only *two* issues are possible ... and a compromise is no longer possible'. Now it is entirely correct and helpful to examine the case of two issues. The point is, however, that *you are then discussing something other than PR.* Carroll was making a mistake rather in the nature of a Freudian slip and the object of his

29 Cf. Collingwood, *The Life and Letters of Lewis Carroll*, p. 72.

thought was some topic other than PR. The analytical technique he was using was appropriate to a personal choice between two objects of two courses of action. While this leaves open what these objects might be, with the earlier connection of the technique with Alice Liddell, it would at least not be surprising if the object about which Carroll was really thinking was again one in the realm of his love life.

Part of his third letter to the *Gazette* and all of his rather trivial fourth letter, and, at a later date, the *Supplement* and *Postscript to the Supplement*, are probably best regarded as regarded as dealing strictly with the problem of PR rather than as having any connection with a choice of a personal kind with which Carroll might be confronted.

The booklet itself uses the technique of the two-person zero-sum game, in which any benefit for the one side is matched by a loss for the other. It would fit the political problem and also that of personal choice, the two problems being isomorphic - which a presumption from the bulk of his literary work and his theory of committees, that in the booklet some personal choice is one of the objects under consideration.

Edith Denman. If, as the analytical might incline us to believe, Carroll was in love at the time, it could only have been with Edith Denman. As children, Edith and her sister Grace had been much admired by Carroll and he had been delighted to be able to photograph them.[30] At the time we are considering, Carroll was taking up friendships with girls in their teens and Edith Denman was a little older than this. she was the daughter of George Denman, one-time Fellow of Trinity, member of parliament and Judge of the High Court, and her grandfather Thomas Denman had been Lord Chief Justice of England and the first Lord Denman. The family was not dissimilar to the Liddells in social and intellectual distinction, a background that Carroll was well able to appreciate. It is easy to imagine that Edith was now a woman of charm and beauty, and she had also some capability as an artist; Derek Hudson mentions a gift she made to Carroll of one of her oil paintings late in 1881 and which he kept in his rooms at Eastbourne but latterly gave away.[31] The tone of the diaries is that Carroll was able to renew his acquaintance with members of the Denman family easily and without strain; at no time were the meetings frequent, yet they leave some impression of a feeling of well-being and comfort.

We now present the entries in the published Diaries which mention Edith Denman and consider another letter on political parties which Carroll had written.

> May 12 (Sun). 1878. [In London] ... Fell in with Judge Denman, who took me home with him, where I renewed my acquaintance with Mrs. Denman, Edith and Grace (who are as nice as when I met them as children fourteen

---

30 The photograph of Grace is plate 39 in Helmut Gernsheim's *Lewis Carroll Photographer* (London, Max Parrish, 1949).

31 Cf. Hudson, *Lewis Carroll*, pp. ix--x.

> years ago), and saw some very interesting drawings by Edith, whose taste is the same as mine, figure drawing.
>
> April 29 (F). 1881. Finished and sent off letter {'Purity of Election'} to *The St James's Gazette.* Made an expedition down the river side to pick fritillaries for Edith Denman; and sent her a bunch in a tin box, and wrapped in wet cotton-wool she wants to paint them.

The opening sentences of 'Purity of Elections' are by way of a Carrollian quip.

> SIR, - Utopia is a pleasant and well-ordered country, and enjoys many blessings to which our little island is a stranger. Some of these must, no doubt, be by us eternally despaired of (for example, no one is ever bored at a Utopian dinner-party or overcharged by a Utopian cab-driver). Others we may hope with fitting effort to make our own; and among these attainable prizes none seems more precious than 'purity of election'. Utopian electors ... are all sufficiently educated to be able to form independent opinions on the political questions of the day; and in accordance with these opinions they vote without fear or favour. Who dares deny that this is a state of things to be wished for and striven for; and that, even though the jealous Parcae may withhold its full fruition, still the more nearly we can attain to it the better and happier we shall be? This, then, being our goal, what are the main obstacles that beset our path - the primary well-springs of corrupt voting?[32]

We get some help in understanding this from the later diary entry:

> May 5 (M). 1884. Wrote to Spooner (who had invited me to dine) to beg off, on ground that in my old age, I find dinner parties more and more fatiguing. This is quite a new 'departure' - I much grudge giving an evening (even if it were not tiring) to bandying small talk with dull people.

A possible interpretation of the beginning of the letter to the *Gazette* would be that we don't live in Utopia, but in the world of hard facts, and one of these is that Carroll is feeling old and knows that his resources of interest, intellect and energy must be carefully expended.

Somewhere in the background (though this is admittedly very doubtful) may lurk the argument that just as in politics there are corrupt institutions, so in personal choice there are misdirected conventions. Choice should be made without

---

32 Much of 'Purity of Election' is quoted in Collingwood, *The Life and Letters of Lewis Carroll*, pp. 232--4. Carroll had fifty copies of the letter handsomely printed as a single sheet; cf. *The Lewis Carroll Handbook* (ed. Green), p. 103. A copy of the sheet is available in the Bodleian [Library, Oxford].

'fear or favour', irrespective of the labels that society may attach; to be diverted from satisfying one's genuinely felt desires by a misguided convention is tantamount to making a corrupt choice. But when we press the interpretation thus far we may go beyond anything that was in Carroll's mind and we would place only a very weak reliance on it.

Fritillaries for Edith Denman. The conjunction of the despatch of the letter to the *Gazette* and the gift of fritillaries is probably not accidental, and the diary entry is unusually elaborate. At the risk of over-documentation, for some readers may not have a botany book at hand, we would explain that the flower is of the lily family and is fairly common near Oxford and grows in the Christ Church meadows. It blooms for only a few days. 'When the flower-stalk emerges it has its bud right over and lying head downwards close beside it. As the bud enlarges and moves a little space away, it is remarkably like a snake's head, and hence a common name of the flower - 'Snake's Head'. Still drooping, it opens and shows its petals chequered with colour ...; red, black, purple, the colour is produced in ... quaint squares and blotches .... The genetic name, Fritillaria, is from the Latin *fritillus*, a dice-box, because, say some, the chequered appearance is like a dice-board; others, however, say the shape of the flower suggests the dice-box. Probably a confusion of the two ideas is responsible for it'.[33]

If the flower did have any significance for Carroll, it might be through its connection with the chessboard of dice-box, or more probably[34] through its folk-lore in which, among other things, it is connected with coquetry.[35]

His acquaintance with Edith Denman continued:

> Dec. 31 (Sat). 1881. the play of *Patience* will do well for the first experiment which Henrietta [his sister] wishes to make in play-going; and Edith Denman is ready to come with us.

The visit to Alfreton (September-October 1884). In the summer there had been excursions to London, visits with child-friends to the theatre, and Carroll spent his holiday at Eastbourne taking with him as usual literary and other work. Edith

---

33 H Essenhigh Corke and G Clarke Nuttall, *Wild Flowers as they Grow* (1911), vol. iv, pp. 20--21. The German popular name is Schachblume, 'chess flower'.

34 Cf. Bowman, *The Story of Lewis Carroll*, p. 74. Cf. also Collingwood, *The Life and Letters of Lewis Carroll*, pp. 150--1, and *Through the Looking-glass*, chapter ii, in Martin Gardner, *The Annotated Alice* (New York, Clarkson N Potter, 1960).

35 Cf. Richard Folkhard, *Plant Lore, Legends, and Lyrics*, 2nd ed. (London, Sampson, Low, Marston, Searle and Rivington, 1892), p. 188.

Denman had by now married the Rev. William Henry Draper,[36] Vicar of Alfreton, Derbyshire; and Carroll had arranged to visit them and probably had promised to give a talk to Mr. Draper's parishioners.

> Sept. 22 (M). 1884. ... to London, on my way to Alfreton to visit Edith and Mr. Draper. Mr. Draper met me at the station and waked up with me, about a mile. At night an ague-like 'cold fit' attacked me.
>
> Sept. 23 (Tu.). 1884. Worked at my article for the *St. James's* on 'Redistribution'. Edith had in two nice children to see me .... But the ague got so much worse that I let Edith send for Mr. Beddard, their doctor, to come over tomorrow from Nottingham.
>
> Sept. 24 (W). 1884. Stayed in and wrote at article. Mr. Beddard gave my lungs a thorough examination and pronounced me 'a thoroughly healthy man', and that I have a feverish cold of ague-type.
>
> Sept. 25 (Th). 1884. A Canon Hole called: he is a friend of Tenniel and the *Punch* staff.
>
> Oct. 4 (Sat). 1884. Instead of the three days I meant to stay, I have inflicted by self for twelve!
>
> Oct. 6 (M). 1884. Sent the St James's my article on Redistribution.
>
> Oct. 11 (Sat). 1884. It appeared.
>
> Oct. 13 (M). 1884. To Eastbourne.
>
> Oct. 16 (Th). 1884. Eastbourne to Christ Church - nearly a week late.
>
> Oct. 28 (Tu.). 1884. Passed for 'Press' the first sheet of the pamphlet [*The Principles of Parliamentary Representation*].

We have no information as to whether Carroll had worked on proportional representation during the vacation, but he was now acquainted with some of the work of the Cambridge School and it enabled him to express his thoughts and feelings more connectedly. No sooner had he reached Alfreton than he was struck by some kind of fever. He found his thoughts leaping to make new connections and to arrive at new generalisations in some kind of heightened mental activity. He had suddenly acquired an inestimable ally, but one whose services would be available to

---

36 Afterwards Master of the Temple. For a biographical sketch see *The Times*, 11 Aug. 1933.

him for only a short time; change his abode and the fever might vanish overnight. Recognizing the advantages of his situation, Carroll used it to the full; he prolonged the visit to about a fortnight and did not leave until he had given shape and structure to the thoughts forming in his mind. He was writing about proportional representation but also, at one remove, about the problem isomorphic with this, the division in his own nature created by his love for Edith Denman. Just as his earlier theory was a way of symbolising his thoughts and feelings in connection with Alice Liddell and making their interaction more intelligible, so *The Principles of Parliamentary Representation* was both a theory of PR and a way of symbolising the desires and fears awakened by Edith Denman and a way of making them in some degree amenable to the processes of rational thought.

At Alfreton he gave the talk to the congregation he had promised and Mr. Draper afterwards published it[37] with a preface explaining the circumstances.

> The writer well remembers his nervous, highly-strung manner as he stood before the little room full of simple people, few of whom had any idea of the world-wide reputation of that shy, slight figure before them.
>
> When the lecture was over, he handed the manuscript to me, saying 'Do what you like with it' ...
>
> It remains to add one or two more associations that cling to it and make the remembrance more vivid still. When Lewis Carroll was staying in the house, there came to call a certain genial and by no means shy Dean [Canon Hole], who, without realizing what he was doing, proceeded, in the presence of other callers, to make some remark identifying Mr. Dodgson as the author of his books. There followed an immense explosion immediately on the visitor's departure, with a pathetic request that, if there were any risk of a repetition of the call, due warning might be given, and the retreat secured.[38]

For being disclosed as the author of Alice, the outburst seems overdone. It was more the start of one who felt guilty in some quite other matter. His lecture is available and it would be interesting to have the pronouncements of the psycho-analysists on Carroll's state of mind at the time.

In the article 'Redistribution' that he wrote at Alfreton, his theory of PR is fully worked out. Getting the booklet from this and his earlier contributions to The St. James's Gazette must have been a job for scissors and paste, with the insertion of introductory and connecting passages.

---

37 Lewis Carroll, *Feeding the Mind* (1907, pp. i--xii and 15--31), with a prefactory note by William H. Draper, reprinted by Roger Lancelyn Green (ed.), *The Works of Lewis Carroll*, pp. 1071--4. [See also S H Goodacre, *Feeding the Mind: a centenary celebration of Lewis Carroll's visit to Alfreton in 1884* (privately printed for the author by Parker & Son, Burton on Trent, 1984).]

38 *Op. Cit.* pp. v--vii.

Just a little after this Carroll missed two notices in *The Times*, one announcing the birth of a son to Edith Denman on 15 December and the other a notice of her death on 30 December, aged twenty-nine years.

Jan. 5 (M). 1885. Heard of the death of my dear old friend Edith Draper.

Conclusion. It had seemed to me, on the grounds I have mentioned, that Carroll's booklet was the outcome of various circumstances, one of which was his love for Edith Denman, and that without this the booklet would never have been written. some corroboration of this view was provided in an unexpected way. The Rev. Mr. Draper married again and a daughter of the second marriage, Lady Hestor Armstrong, has been able to fill in some of the gaps in the history.

Carroll, we know, was interested in beautiful young women, provided there was no danger of marriage in the background; and this was the case with Edith Denman at the time. She was firmly determined to marry a handsome young clergyman of her own age who was then a student at Keble College, Oxford. There was strong opposition from her family and the issue of the affair, though the couple were deeply committed to each other, had not been finally settled. Her circumstances were sufficiently ambiguous, without a formal engagement and with very strong opposition from her family, for marriage with some other to be possible. for Carroll, the young clergyman in the background had a reassuring effect: there was always the possibility of a gentlemanly surrender. Thus insured against the dire event of marriage Carroll, we conjecture, ventured to allow himself to fall in love with Edith Denman. He debated within himself whether to keep to the daily round at Christ Church and the life of the scholar, enlivened by visits to the theatre with children and his friendships with older girls, or to abandon this and seek a marriage which seemed unusually attractive. While the outcome may never have been in real doubt, Carroll pondered the issue; and his mind symbolised the conflict between the two sides of his nature, the two sets of opposing desires, as the contest between two political parties, with the need for a rational honest choice.

At the parental insistence, the young couple agreed for a considerable time not to see each other; but there was an accidental meeting at Preston railway station. They sat and talked things over and, when they announced their firm determination to marry, parental disapproval gave way and their home became the vicarage where later Carroll was to be their guest. Lady Armstrong remembers that for long Edith Draper's photograph occupied a place of honour on the table of the drawing room in their household. Edith's son, Mark Draper, exceedingly dark with blue eyes, very good-looking and happy-go-lucky, was first articled to a solicitor, then became an actor, later a photographer, and latterly had his own theatrical company and enjoyed some success as a producer of plays. He was killed in 1916 on a training flight a fortnight after he had joined the Royal Air Force and was buried beside his mother in the churchyard at Alfreton.

When she died at the age of twenty-nine Edith Denman deserved a more truthful epitaph than 'Heard of the death of my dear old friend Edith Draper'.

Equally incongruous was the tribute paid to Carroll's booklet. Dismissed with no more than a few deprecatory remarks by a single author, the most distinguished contribution that has been made to Political Science since the seventeenth century was allowed to rest in peace.

Part 3. An Analysis of Carroll's Argument

3.1. Carroll's Scheme of Proportional Representation

Scope. From almost the first page to the last, Lewis Carroll's booklet *The Principles of Parliamentary Representation* (or *PPR* as we shall refer to it) is concerned with the Droop quota - and the novelty is that the quota is being used within a two-party system. The present chapter considers the use of the Droop quota in general and more particularly within the two-party system, and has the general aim of making Carroll's booklet more intelligible. We start with a description of the scheme of Proportional Representation (PR) which he recommends for use in England at the time he is writing; and the theory we go on to develop will answer some of the questions arising directly out of Carroll's scheme.

A proposal based on party. In the polity which Carroll envisages, there are two political parties, A and B, say; and for the purposes of arriving at a model, he assumes that each elector will have a particular shape of 'preference curve'. The elector who supports the party A is taken to prefer any A-candidate to any B-candidate, and the supporter of party B prefers any B-candidate to any A-candidate. He does not, of course, state his supposition in these terms, but each of the four arithmetical examples given in *PPR* makes it clear that this is his view of the preferences of the electors. Party consideration is assumed to dominate, and to swamp any difference of a personal kind that the voter may find among the candidates.

The constituency will return say, four or five members, and the elector will be entitled to cast only a single vote. When the poll is closed, each candidate getting a Droop quota will be elected. But as a rule any candidate elected will have votes surplus to his needs, that is more than a Droop quota, and other candidates will have fewer votes than entitle them to election. Carroll's suggestion is that the candidates of a party shall meet and exchange surplus votes of these two kinds, so that the party shall take up all the seats to which it is entitled, by the total number of votes cast for its candidates.

> The Elector must understand that, in giving his vote to [a Candidate] ... he gives it to him as his absolute property, to use for himself, or to transfer to other candidates, or to leave unused. If he cannot trust the man, for whom he votes, so far as to believe that he will use the vote for the best, how comes it that he can trust him so far as to wish to return him as a Member?
>
> [Let the returning-officer] announce as "returned" any Candidate who has received the quota needed to return *one*. If there are still Members to return, let him appoint a time and place for all the Candidates to appear before him; and any two or more Candidates may then formally signify that they wish their votes to be clubbed together, and may nominate so many of themselves as can be returned by the votes so clubbed. They must of course include in their nomination any of themselves who have been already declared to be returned.
>
> ... This method would enable each of the parties in a District to return as many Members as it could muster the proper quota for, no matter how the votes were distributed. There would be no risk of a seat being left vacant through rivalry between two Candidates of the same party: an unwritten law would soon come to be recognised - that the one with fewest votes should give way. With Candidates of two opposite parties, such a difficulty could not arise at all: one or other of them could always be returned by the surplus votes of his own party. The only exception to this would be the occurrence (a very rare one) of an exact balance of votes. This might happen, even in the case of a single-Member constituency, if each of two Candidates got exactly half the votes. Of course, in such a case, somebody must give a casting vote.[1]

Carroll goes on to argue that a scheme of this kind is preferable to the single transferable vote (s.t.v.), which was advocated by the Proportional Representation Society at the time, and had a wide volume of support in parliament and the country at large. We might examine the grounds on which he reaches this conclusion, or the relation of Carroll's scheme to some proposals which had been made by others. But, as we have explained, our aim in the present essay is rather to make Carroll's booklet and his theoretical model intelligible, and answering questions of this kind now, might act as a distraction. The procedure he proposes makes use of the Droop quota. How does the Droop quota operate in a two-party system? Would the parties between them be certain to fill precisely the number of seats available? Would the outcome be fair as between larger parties and smaller parties? To what extent would it achieve PR? These questions would seem to be logically prior to any comparison which might be made between Carroll's scheme

1 Cf. Reprint below, pp. 165--7.

and the s.t.v., or the schemes of others, and it is with finding answers to them that we will be concerned.

3.2. The Desiderata: In at the Deep End

One way of teaching a child to swim is to cast him in at the deep end, but although instances of success have been recorded, it is not a method which has won much favour. No more is it to be recommended in writing a booklet on PR, and Carroll's initial chapter is likely to leave the reader looking for support and unable to find it. Carroll's introduction to the booklet might better have been placed at the end of his argument.

We can offer a few notes by way of elucidation and point some of the pitfalls, but our real attempt to make the booklet intelligible must take the form of successive explanations of parts of the theory. At the same time we can point to the relation of his work to that of Walter Baily and James Garth Marshall, in which his line of thought can be seen in its simpler beginnings.

In a party system, a definition of PR must presuppose that the voters' preference curves are of a certain shape. As one of his *desiderata* Carroll states:

> (4) That the proportions of political parties in the House should be, as nearly as possible, the same as in the whole body of Electors.

This *desideratum* amounts to a definition of PR, and one which Carroll intends to be used only in the context of a two-party system. No use can be made of the definition, however, unless we are able to identify whether a voter or elector belongs to say, the party A or the party B; and this is an aspect of which Carroll is fully aware. Whether the voter or elector belongs to one or other of the two parties will depend on the shape of his preference curve; and it may be well to distinguish the two types of preference curves which enable the definition to be put to use.

At the four different places in his text at which he gives arithmetic examples of elections, the preferences of the voters conform to the type of Fig. 3.1. The reader is probably familiar with diagrams of this kind, and we will mention only that each real point on the horizontal axis is to be taken to denote a candidate, and in Fig. 3.1 the A-candidates are taken to be towards the left-hand end of the axis, while the vertical axis shows the order of preference of a particular voter in regard to the candidates. The characteristic which defines the voter as belonging to party A, say, is that, as in Fig. 3.1, he will rank *any* candidate of party A, higher than *any* candidate of party B. Vice versa, the supporter of party B will rank *any* B-candidate at a higher level of preference than *any* A-candidate. If and only if a preference diagram conforms to this description can we identify the Elector or voter in question as belonging to one or other of the two parties. In the problem with which Carroll's booklet is concerned, this is the only type of diagram which is admissible.

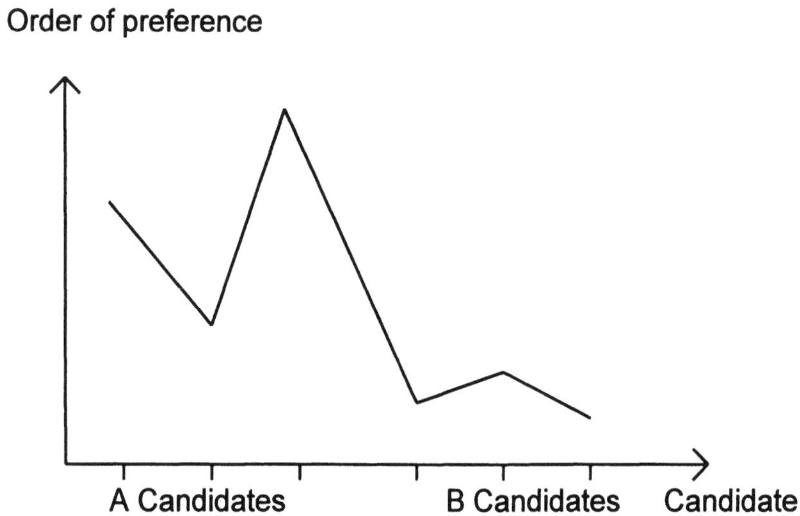

Figure 3.1

As a special sub-class of this type of diagram, it might be that the A-supporter was indifferent as between all the A-candidates who stand in his constituency, and indifferent also as between all the B-candidates. His preference curve would then consist of two plateaus, a higher plateau corresponding to his level of preference for the A-candidates in the election, and a lower plateau corresponding to the B-candidates, as in Fig. 3.2. The theory would work out more simply for this type of preference curve, than for that of Fig. 3.1.

In the problem of PR the data can be stated as the set of preference curves of the voters or electors in the particular constituency or the country at large, as the case may be. Granted that the preference curve of each individual is of the type of Fig. 3.1, or the sub-class Fig. 3.2, each individual can be identified as belonging to one or other of the two parties. A condition of complete PR is then achieved, if and only if the ratio of the number of seats held by A and B is the same as the ratio between their numbers of supporters.

In fact the number of seats, whether for the constituency or the country at large, will be small by comparison with the number of voters, and seats are not divisible but must be awarded as a unit at a time; and it will rarely be possible for anything more than an approximation to be reached between the ratio of seats and the ratio of numbers of voters. The more fully is equality reached between these two ratios, the more fully will PR be achieved.

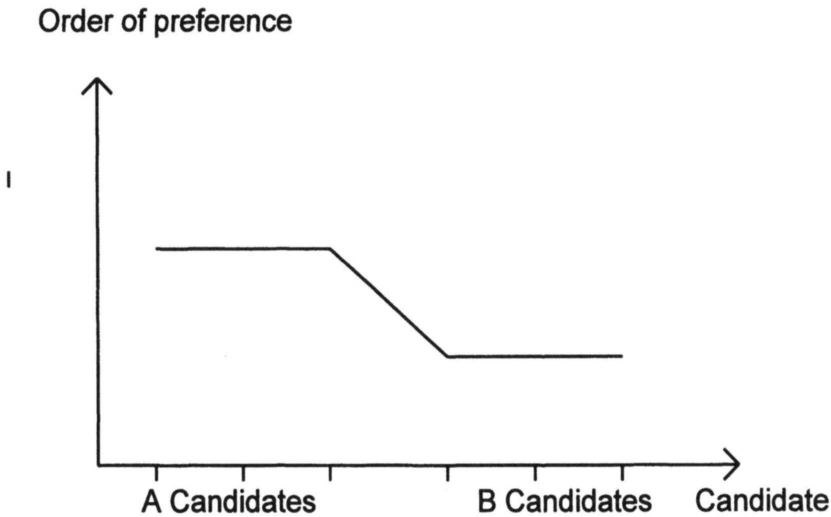

Figure 3.2

Let A = number of A-voters and A* = number of A-seats, and B = number of B-voters and B* = number of B-seats. Then whether for the constituency or the country at large, PR is fully attained if and only if $\frac{A^*}{B^*} = \frac{A}{B}$, and therefore, $\frac{A^*}{A} = \frac{B^*}{B}$, and the aim of PR is to achieve so far as possible, equality between the two ratios.[2]

Thus given preference curves of the appropriate shape, *desideratum* (4) provides both a definition of the *meaning* of PR, and the basis of a *criterion* of the extent to which PR is realized in any given circumstances. It will be found, however, that the booklet accepts both a meaning and a criterion of PR which is different from this, being cast in terms of 'the number of wasted votes' and 'the number of voters unrepresented', and the booklet makes no effort to measure the extent to which PR is achieved, along the lines of these ratios.

The use of undefined terms. Some of the other *desiderata* in the chapter are of a concrete nature and need no comment:

(5) That the process of voting should be as simple as possible.

---

2 Cf. E V Huntington, ['The apportionment of Representatives in Congress', *Transactions of the American Mathematical Society* 30: (1928), 85--110; 'A paradox in the scoring of competing teams', *Science* 88 (1938), no. 2282 (September 23), 287--8.]

(6) That the process of counting the votes, and announcing the result, should be as simple as possible.

But the other abstract *desiderata* which Carroll goes on to formulate have the serious fault that they employ terms of a seemingly common-sense nature, but in reality with a special connotation only to be given later in the text, *Desiderata* (1), (2), (3), (7) involve 'the number of electors represented, or unrepresented', and 'the number of wasted votes'. Each of these terms is later given a nominal definition which, in reading chapter I[3], the reader can only remotely guess at.

Again *desideratum* (4) discussed above, is referred to at no point in the text and is given no explicit role in Carroll's subsequent argument. Yet the whole of his argument turns on this *desideratum*.

Before attempting to introduce *desiderata* or criteria, the treatment given below will present the elements of a positive theory of a procedure using the Droop quota, this being the procedure on which Carroll will rely to secure PR. A positive theory will show the objectives which it is possible to achieve and is a necessary preliminary to any normative theory.

A spurious principle. The third in Carroll's list of *desiderata* states the principle:

(3) That the number of unrepresented Electors should be as small as possible.

Sometimes in the text, Carroll states the same principle as being 'that the number of unused votes' should be as small as possible, equating 'the number of unused votes' and 'the number of electors unrepresented'. Each of these phrases is a technical term, whose meaning at the commencement of the booklet, the reader cannot possibly understand, or even remotely guess at.

Now any further principle will be admissible, only if it is logically equivalent to the principle of PR, or at the very least does not run counter to it; and it is shown in our text that, giving the phrase the meaning Carroll assigns to it, the connection between 'the number of electors unrepresented' and the principle of PR is exceedingly complicated. The reader is advised to disregard all mention of electors being 'unrepresented' in Carroll's opening chapter.

The question asked. *The Principles of Parliamentary Representation* is concerned solely with the two-party regime. Carroll speaks of them as the 'red' and 'blue' parties, and we will refer to them as the parties A and B. Where the context makes the meaning clear, we will also use A to denote the number of A-supporters, and B the number of B-supporters, as in A = 500, B = 400. Every elector in the community is assumed to fall into one or other of these two parties.

---

3 [Reprint below, p. 152].

A further simplification made by Carroll, is that he is concerned with only a very limited range of procedures. He makes the basis of his enquiry the m-member constituency ($m \geq 1$), in which the available seats will be awarded to the m candidates getting the highest number of votes - but where provision is made for the party being able to distribute all the votes got for its candidates, in such a way as to fill the maximum number of seats. If, for instance, A = 600, it will be assumed that the party A is able to give 300 votes to each of two candidates, or 200 votes to each of three candidates, according as one or other of these arrangements will maximize the number of seats that it fills.

A contest between the parties having these features may be conducted under various electoral procedures which give the elector more votes or fewer votes; and Carroll envisages that the contest in the m-member constituency may be held under any of m different procedures. The elector may be given one vote, or two votes, ..., or m votes, subject always to the condition that he can give not more than one vote to any particular candidate. For example in the five-member constituency, under one electoral procedure the elector may have five votes, to be given to five different candidates; under another of the procedures envisaged as possible, he could be given three votes to be cast for three different candidates; or he might have only a single vote.

The question which Carroll sets out to answer is: Which of this set of m procedures is the most suitable? or again, As between any pair of the procedures we have defined, which is the more suitable of the two?

The antecedent value judgment. It is, therefore, on a normative enquiry that we have embarked, and the reader is likely to be aware of the complications which, in Economics, attend any enquiry of this kind. The economist who is reporting to a committee about say, some proposed fiscal changes, or to the board of directors of a firm in respect of certain proposed developments, will be expected to point out the economic implications of the changes which they propose. And the members of the committee may accept in full his findings about the consequences of these changes: but even so, it must be left to the individual committee member to form his own assessment of the *relative desirability* of the possible changes. In the last analysis the so-called 'value judgment' of the individual is called upon, when he forms an opinion about which changes are more desirable, and which less desirable. This is not a matter on which the technical competence of the economist enables him to pronounce. It involves more than economic reasoning; and the same diagnosis in purely economic terms, is compatible with different rankings of the changes in order of preference, by the different members of the committee.[4]

The theory of PR is in a quite different case from this. Let us take it that a decision has already been reached as to which classes in the community are to

---

4 Cf. Lional Robbins (Lord Robbins), *An Essay on the Nature and Significance of Economic Science*, 2nd ed., (London, Macmillan, 1952), pp. 24--28.

receive the franchise, say all citizens over the age of 21, or over the age of 18, and that the owners of business premises are to be given a double vote, and so on. Then by the very fact that the enquiry we conduct is an enquiry in PR, it is accepted that the objective is to give the various classes who enjoy the franchise, representation in proportion to their numbers. This is the only value judgment involved, and it lies outside the enquiry into the electoral arrangements by which PR will be achieved. The value judgment and decision to make the enquiry, is anterior to the investigation of the electoral arrangements by which the objective of PR will be achieved. The desirability of securing PR has been accepted, before the investigation begins.

3.3. The Droop Quota in a Two-Party System

Introductory. During the nineteenth century much of the theorizing on PR took the form of trying to find a suitable electoral quota, and in England the view came to be generally held that the Droop quota was more suitable than any other. Later we will cite the opinions of this quota held by some of the main writers of the period. The present investigation will consider the operation of the Droop quota in the simplest possible set of conditions, that in which there are only two political parties. In this we follow the example given by Lewis Carroll, and at an earlier date by James Garth Marshall.

The objective of PR is to make the ratio of A-representatives to B-representatives as nearly as possible the same as the ratio of A-supporters to B-supporters. The wider problem of PR is concerned with these two ratios in the chamber as a whole, and the narrower problem with making the two ratios as nearly as possible equal in the single election in the particular constituency. It is only with the problem of the constituency that Carroll or the present writer is concerned. The remaining algebraic symbols we employ will be those of *The Principles of Parliamentary Representation*.

e = the total number of voters in the constituency under consideration, $e \geq 2$.

Denoting by A the number of A-supporters and by B the number of B-supporters, the party A must contain at least some A-supporter and the party B at least one B-supporter. The possible composition of e, therefore, is given by $1 \leq A \leq e - 1$, $1 \leq B \leq e - 1$, $A + B = e$.

m = the number of seats in the constituency under consideration.

v = the (integer) number of votes allowed to the elector, and the permissible range of v is $v = 1$ or 2 ... or m; but it is stipulated that if the procedure concerned allows the elector more than one vote, the elector shall give no more than one vote to a single candidate.

Each value of v defines a distinct electoral procedure; and Carroll's objective is to examine the degree of PR that is provided by the different procedures got by giving the different possible values to v. In fact, when he has completed his investigation he comes down quite unambiguously in favour of the procedure which takes $v = 1$ and gives the elector only a single vote. We ourselves fully agree with this view which we arrive at by a different route from that chosen by Carroll. Our own treatment of the procedure for which $v = 1$, is largely of a diagrammatic nature; and it has the advantage of being readily extensible to all the other procedures in the family, given by making $v = 2$, $v = 3$, ..., $v = m$.

The Droop quota. The Droop quota specifies the number of votes necessary and sufficient for a candidate to secure election. Droop designates this number of votes so as to achieve the following objectives:

(i) *With m seats available in the constituency, the size of the quota shall avoid the possibility of electing as many as* $(m + 1)$ *candidates.*

Denoting the size of the Droop quota by Q_d, a sufficient condition to achieve this objective is that

$$(m+1)Q_d > e$$

$$Q_d > \frac{e}{m+1}$$

(ii) *The quota shall be a positive integer number of votes.*

The only values satisfying the requirement $Q_d > e/(m+1)$ and making Q_d a whole number, are

$$\frac{e}{m+1}+i, \quad 0<i\le 1,$$

where i is the fraction, or possibly unity, needed to make up the quantity $e/(m+1)$ to the integer next above $e/(m+1)$ in size, together with the further integers

$$\frac{e}{m+1}+i+1, \quad \frac{e}{m+1}+i+2, \quad \ldots$$

(iii) *The Droop quota shall fill as many seats as possible, short of* $(m + 1)$.

This is equivalent to asserting that the Droop quota shall be the smallest in the series of integers just named; which gives

$$Q_d = \frac{e}{m+1}+i, \quad 0<i\le \ ,$$

i, the fraction involved, or possibly unity, must be expressible in the form which has $(m + 1)$ for denominator; so that in general $i = t/(m+1)$, where t is an integer such that $1 \le t \le m + 1$. This gives

$$Q_d = \frac{e}{m+1} + \frac{t}{m+1}$$
$$= \frac{e+t}{m+1}, \quad 0 \leq t \leq m+1$$

This will be the formula which we nearly always employ for the Droop quota.

J G Marshall. The definitions of pronouncements of the main writers on the Droop quota differ scarcely at all. J G Marshall, whose work we discuss in a separate section, avoided defining *any* concept, the Droop quota along with the rest, and allowed his theory to take shape in numerical examples. From his use of the notion in these examples, he must be considered to be the true originator of the 'Droop' quota.

H R Droop. Droop's proposal for the use of the quota which came to bear his name, was made in a pamphlet[5] of 1868. In a paper of a slightly later date he gives this summary view of the quota:

> For instance in a constituency electing five representatives, any fraction exceeding one sixth of the voters could ... secure one representative, and if they amounted to two-sixths or three-sixths of the voters, they could obtain their proportion of two or three representatives.[6]

Droop's main theoretical discussion is in the paper 'On Methods of Electing Representatives', *Journal of the Statistical Society*, June 1881. Replacing the algebraic symbols by those used by Carroll, his theoretical argument, contracted a little, would run:

Take the m-member constituency in which the elector is allowed a single vote, and suppose that in all e votes are cast. Let $\left[e/(m+1)\right]+i$ be the next whole number greater than $e/(m+1)$, so that $0 < i \leq 1$. Then $\left[e/(m+1)\right]+i$ votes will properly be considered a sufficient number of votes to elect one representative. For if $\left[e/(m+1)\right]+i$ votes be given for each of m candidates, the votes remaining undisposed of will be

$$e - m\left(\frac{e}{m+1} + i\right) = \frac{e}{m+1} - mi$$

---

5 H R Droop, *On Methods of Electing Representatives* (London, Macmillan, 1868), p. 32.

6 Cf. H R Droop, *Proportional Representation as Applied to the Elections of Local Governing Bodies* (London, Wildy and Sons, 1871), p. 3.

which is manifestly less than $[e/(m+1)]+i$, and therefore the votes remaining undisposed of would not displace any candidate who had obtained $[e/(m+1)]+i$ votes.[7]

Walter Baily. In the second of his two pamphlets on PR, Baily says:

> The number of votes to be retained for a candidate must be enough to make his election certain, whatever combination may be made of the other votes given in the election; *the smallest number which will suffice for this is the true Quota* ... There is no difficulty in finding this number ... Divide the number of votes by the number just above that of the members to be elected, and take as the quota the number just above the Quotient.[8]

The common sense behind this choice of the quota as the number *just above* the quotient, is that while the quota shall not fill as many as $(m + 1)$ seats in our notation, it will either fill the m seats available or at any rate come as near as possible doing so.

Lewis Carroll. The theory of the Droop quota given by Carroll is enmeshed with his view of the number of electors represented by a member for the constituency, and we will not attempt to disentangle the two. At the same time this quotation will make it clear that his symbols echo the words of Baily.

> Let us call [the] necessary and sufficient quota 'Q'.
>
> Now, in order that Q may be *sufficient*, it must not be possible for m other Candidates to obtain Q votes each; i.e. $(m+1)Q$ must be greater than e; i.e. Q must be greater than $e/(m+1)$. Also, in order that Q may be *necessary*, it must be the whole number greater than the fraction. Hence approximately $Q = e/(m+1)$.[9]

In another connection later in the booklet, Carroll explains that he will mean by the 'necessary number of votes', the number of votes only just large enough for the purpose; and this again is a clear reference to Baily's position.

---

7 H R Droop, 'On Methods of Electing Representatives', *Journal of the Statistical Society*, June 1881, pp. 172--3.

8 Cf. Walter Baily, *Proportional Representation in Large Constituencies,* (London, Ridgway, 1879), pp. 6--7, our italics.

9 Cf. Reprint below, p. 154.

Given e and m and $e \geq mt$, the size of the Droop quota is unique and definite. Proof is that for any given values of e and m, $e/(m+1)$ has a definite size, whilst t, the fraction of a vote or possibly the single vote which must be added to give the whole number of votes just greater than $e/(m+1)$, is also of a definite size.

Using the formula $Q_d = (e+t)/(m+1)$, $1 \leq t \leq m+1$:

For $e = 7$, $m = 7$, $Q_d = \frac{7}{8} + \frac{1}{8} = 1$, $t = 1$; and $e = mt$.

For $e = 20$, $m = 3$, $Q_d = \frac{20}{4} + \frac{4}{4} + 4/4 = 6$, $t = 4$; and $e > mt$.

For $e = 1001$, $m = 4$, $Q_d = \frac{1001}{5} + \frac{4}{5} = 201$, $t = 4$; and $e > mt$.

Proposition 1. For $e > mt$ the number of votes cast, e, will be greater than m quotas of size $(e+t)/(m+1)$; for $e = mt$ the total number of votes cast will be exactly m quotas of size $(e+t)/(m+1)$; and for $e < mt$ the total number of votes cast will be less than m quotas of size $(e+t)/(m+1)$.

We may write

$$e = \frac{m(e+t)}{m+1} + \frac{e-mt}{m+1}$$

Now e is a positive integer and so also, by the manner in which it is defined, is the quantity $(e+t)/(m+1)$. It follows that the second term on the right-hand side of this equation is also an integral number of votes, which may be positive, or zero, or negative.

For $e > mt$, the second term is positive and e amounts to more than m quotas of size $(e+t)/(m+1)$.

For $e = mt$, the second term is zero, and e is exactly equal to m quotas of size $(e+t)/(m+1)$.

For $e < mt$, the second term is negative and must therefore denote a negative integral number of votes, so that e amounts to a number of votes which is less than m quotas of size $(e+t)/(m+1)$. We can therefore say, without further proof, that for $e < mt$ the Droop quota does not exist.

Proposition 2. For $e \geq mt$, the total number of votes $e = m.Q_d + (Q_d - t)$ votes.

$$e = m\left(\frac{e+t}{m+1}\right) + \left(\frac{e+t}{m+1} - 1\right), \quad 1 \le t \le m+1$$
$$= mQ_d + \left(Q_d - t\right)\text{votes.}$$

This formula is repeatedly of use in discussions connected with the dot graph below.

For $e = mt$ the second term on the right-hand side of the equation becomes zero. The Droop quota = t and the total number of votes cast becomes m Droop quotas.

In a 2-party system, even with *m* quite small and only *m* candidates standing, it is unlikely that the Droop quota would fill the *m* seats, unless helped by some electoral device such as a list system.

(i) $(m+1)Q_d = e+t$, and at most the Droop quota can fill m seats. Obviously if fewer than m candidates stand, not all the m seats available can be filled. From Proposition 2, if exactly m candidates stand, but the spread of votes among the candidates is uneven, it will be sufficient to prevent the filling of all the m seats available that a single candidate should get as many as two Droop quotas, or that only two candidates should get between them as many as three Droop quotas and so on.

(ii) $e/m = Q_d + (Q_d - t)/m$. The closer is t to the Droop quota in size, the more even will have to be the spread of votes among the m candidates, in order that all the m seats available shall be filled. In the limit for $t = Q_d$, $e = mt$, and only a completely even spread of the votes among the m candidates, would fill all the available seats.

(iii) Looking at the matter in terms of proportions, from the formula just given above, if m candidates stand, the average number of votes, e/m, will amount to less than $(1+1/m)$ Droop quotas per candidate. For $m = 1$ this will amount to less than two Droop quotas per candidate, ..., for $m = 5$, say, it will amount to less than $(6/5)Q_d$ per candidate, and so on. This helps to bring out the difficulty of each of m candidates getting a Droop quota, as the number of seats in the constituency increases.

These considerations show that even in a two-party system in which the parties succeed in restricting the numbers of their candidates so that a total of only m stand, the Droop quota will be unlikely to be able to fill these seats, unless special arrangements are made to that end.

The part played by the indivisibility of seats, in hindering the two parties from filling the *m* seats available. Suppose A = 28, B = 9, $e = 37$, $m = 6$, giving

$Q_d = 6$, $t = 5$ and $e > mt$. A holds 4-4/6 Droop quotas and B 1-3/6 quotas. But the parties between them hold only five *complete* quotas, the one holding four complete quotas and the other one.

It is impossible to award to a party a fraction of a seat, and Carroll stipulates that in his electoral procedures, a party shall be awarded just as many seats as it holds *complete* Droop quotas of votes. In the above example the parties would fill only five seats and not the six seats available in the constituency.

Some supplement to the electoral procedure might be suggested which, in the event of the quota filling only (m - 1) of the available seats, as in this example, would make an appointment to the remaining vacant seat. The obvious suggestion would be to award a seat for each full Droop quota held by a party and then award the remaining seat to the party which comes nearest to having a further full Droop quota. We might have no objection to this course. But our real concern is to understand the working of the Droop quota and to discover the circumstances in which the quota would fail, in the first instance, to make all the m appointments which are desirable, when the stipulation is made that a seat shall be the reward for a *complete* Droop quota.

Outside a party system, the Droop quota could scarcely work. While there is nothing in the concept of the Droop quota which need confine its use to a party system, outside a party system the quota would probably work badly. Suppose, for instance, this quota to be used in a constituency with 300 voters and four seats, giving $Q_d = 61$; and suppose that only five candidates stand, getting say 100, 80, 60, 40 and 20 votes. Then two seats would be filled and the remaining three left empty, with no suggestion as to how the Droop quota might be used to fill these remaining seats.

In these circumstances the Droop quota would seem to be an irrelevance. The votes cast represent the first-preferences of the electors concerned. The simpler electoral arrangement of filling the first place by the candidate with the highest number of votes, the second place by the candidate with the second-highest number of votes, and so on, would at least fill all the available seats and would seem to be preferable to dragging the Droop quota into circumstances where it makes a nonsense.

We might generalize this result and say that outside a party system, use of the Droop quota would seem to be inappropriate.

The bugbear of all first-preference voting, 'the waste of votes'. At the same time, the system of first-preference voting just described, gives almost unlimited scope for waste of votes. As an example take an election in which $e = 300$, $m = 4$, and five candidates stand getting 130, 100, 40, 20 and 10 votes respectively. Then one candidate will be elected with only 40 votes and another with only 20. Walter Baily's summing-up of a situation of this kind is: 'Some votes are wasted by being given to candidates who have enough without them, and others are wasted by being given to candidates who are not elected'.[10]

The Droop quota as a pricing system. A common-sense view may take the Droop quota in a scheme of this kind as being the price of a seat, but with the restriction that a party can buy only a whole seat at a time, and not the fraction of a seat. The quota is made the basis of a pricing system.

One of the objections brought against the pricing system of an economy is that there is no presumption that the initial distribution of income is right and just, and goods may be wrongly distributed. In the present instance, however, no difficulty of this kind arises and there is the presumption that the votes cast express the wishes of the electors, and each party is entitled to just as many votes as it receives.

On the other hand in the economic market the commodities traded are finely divisible and can be purchased in small quantities at a time - whereas it is not open to a party to purchase a fraction of a seat. The political market offers no scope for fine adjustment and the arrangements arrived at may be rough and ready.

As an example, take A with 1977 supporters and B with 4020 supporters, giving $e = 5997$; and with $m = 5$, $Q_d = 1000$.

A gets 1.977 Droop quotas and B gets 4.020 Droop quotas. With 33% of the votes, A gets 20% of the seats, and B with 67% of the votes gets 80% of the seats. It might be thought to be closer proportionally if A were given two seats, that is 40% of the seats, and B three seats, that is 60%.

The maximum discrepancy of this kind occurs when one party gets nearly a full Droop quota of votes over and above its number of complete Droop quotas, and the other party gets few votes or none over and above its number of complete Droop quotas. Also the discrepancy is magnified for the small party which would in any event get only one or two seats. In such circumstances the change of a few votes would give one party one seat more, and the other one seat fewer.[11]

3.4. The Representation of the Droop Quota

The 'Droop.Representation'. Some properties of the quota in a two-party system will now be considered using the dot graph, a device discovered by the Pythagoreans, which continues to be used in number theory.[12] First we show how to construct the representation, given the pair of values (e,m), the number of voters

---

10 Walter Baily, *Proportional Representation in Large Constituencies*, p. 4.

11 The single transferable vote, it should be mentioned, has precisely this same defect through its reliance on the Droop quota.

12 Cf. John Burnet, *Greek Philosophy, Thales to Plato*, (London, Macmillan, 1960), pp. 51--6, and G H Hardy and E M Wright, *An Introduction to the Theory of Numbers*, (Oxford, Clarendon, 1938), chapter on Partitions.

and the number of members for the constituency, provided $e \geq mt$. If $e < mt$, we will show no representation of the kind can be made.

The representation is an array of e dots, each dot to represent a voter, with the dots arranged so as to correspond to the properties of the Droop quota. Knowing e and m and with $e \geq mt$, we can compute the Droop quota from the formula $Q_d = (e+t)/(m+1)$, $1 \leq t \leq m+1$.

With m seats in the constituency, set out m rows, each row containing a Droop quota of dots. This leaves to be taken into account $(e - mQ_d) = (Q_d - t)$ voters, and to represent them, add another row, that is, an $(m + 1)^{\text{th}}$ row in the graph - and place $(Q_d - t)$ dots at the left-hand end of this row. At the right-hand end of this lowest row, add t small crosses, so as to give an arrangement of the type shown in Fig. 3.3.

For the given pair of values (e,m) such that $e \geq mt$, the size of the Droop quota is definite, t is definite, $(Q_d - t)$ is definite. The number of dots to be entered in each of the upper m rows is definite, as is the number of dots to be entered in the lowest row and also the number of crosses t, $1 \leq t \leq m + 1$, by which these dots will be followed. Thus for $e \geq mt$, the Droop representation is unique.

Example with $e > mt$. Choosing $e = 37$, $m = 4$, gives $Q_d = 37/5 + 3/5 = 8$, $t = 3$, $mt = 12$, and $e > mt$. $Q_d - t = 5$.

To get the Droop representation place 8 dots, $(Q_d = 8)$, in each of the $m = 4$ upper rows. At the left-hand end of the $(m + 1)^{\text{th}}$ (i.e. the 5th) row, place the remaining $Q_d - t = 5$ dots to make up the total of $e = 37$ dots, and at the right-hand end of this row place $t = 3$ small crosses. This gives the rectangular array of dots and crosses shown in Fig. 3.3.

Figure 3.3

For $e = mt$, we have shown earlier, $Q_d = m + 1 = t$. In these circumstances the Droop representation will consist of m upper rows each containing $(m + 1)$ dots, together with a lowest, $(m + 1)^{\text{th}}$ row, containing $(Q_d - t) = (m+1) - (m+1) = 0$ dots and $t = (m + 1)$ small crosses; that is, the lowest row will consist of only small crosses and will have just so many crosses as each of the upper rows has dots.

Example with $e = mt$. Taking $e = 12$, $m = 3$, gives $Q_d = 12/4 + 4/4 = 4$, $t = 4$. The Droop representation consists of $(m + 1) = 4$ rows in all, each of the first m rows containing $(m + 1)$ dots, and the lowest row consisting of $(m + 1)$ small crosses. This gives a *square* array of dots and crosses, in which $Q_d = (m+1)$, and the size of the array, including the lowest row, is $(m + 1)\times(m + 1)$.

For $e > mt$, the array will be *rectangular* in shape, with the *possibility of the square array* as a particular case. But where it occurs the square array will differ from that for $e = mt$, in that for $e > mt$ the lowest row will consist of t small crosses, situated after $Q_d - t > 0$ dots at the left-hand end of the lowest row - whereas for $e = mt$ there are no dots in the lowest row and only $Q_d = (m+1)$ small crosses.

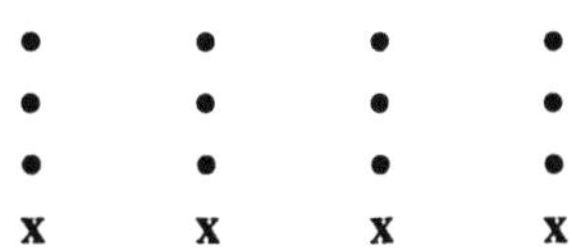

Figure 3.4

For $e > mt$ the narrowest shape of rectangle is given by $e = m$, $(Q_d - t) = m/(m+1) + 1/(m+1) = 1$, $t = 1$. Each of the m upper rows of the graph consists of only a simple dot, and the lowest row of a single cross, as in Fig. 3.5 for $e = m = 5$.

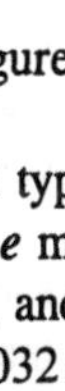

Figure 3.5

At the other extreme a broad type of array corresponds to the so-called 'parliamentary' constituency, in which e may run into thousands, while m is only a small number of seats, say 3 or 4 or 5, and of course $e > mt$. As an instance take $e = 25032$, $m = 4$, giving $Q_d = 25032/5 + 3/5 = 5007$, $t = 3$. A sufficient indication of the Droop representation is given by a graph which has 5 rows in all, while the number of dots in each of the 4 upper rows, 5007, is shown by placing this number itself, 5007, in the topmost row; and likewise the number of dots in the lowest row is shown by placing the number involved, 5004, at the left hand of the row. The lowest row is completed by adding three (t) small crosses at its right-hand end, as in Fig. 3.6.

•	•	5,007	•	•	•	•
•	•	5,007	•	•	•	•
•	•	5,007	•	•	•	•
•	•	5,007	•	•	•	•
•	•	5,004	•	x	x	x

Figure 3.6

Two sets of propositions, the one algebraic and the other graphical, which are logically equivalent. Suppose we start off knowing that the 'quota' in an electoral procedure is the number of votes which is legally prescribed for the election of a candidate, but that we have no knowledge of the particular type of quota known as the 'Droop quota'. Consider a constituency of e voters, each casting a simple vote, to fill m seats. And let us suppose that we find it possible to arrange e dots in such a way that they form say, m rows ($m \geq 1$), each containing the same number of dots, X dots say, together with a lowest row which either contains some dots or is possibly empty, but which contains between one and $(m + 1)$ dots fewer than each of the upper m rows. The number of dots in the lowest row can be denoted by say, $(X - t^*)$, where $X \geq t^*$, $1 \leq t^* \leq (m + 1)$.

The dots in the lowest row may be arranged in any manner we choose. Suppose we place the $(X - t^*)$ dots at the left-hand end of the row, followed by t^* small crosses, after the manner shown in the above dot graphs. The lowest row will the contain X dots on the left and t^* small crosses on the right.

Now if we choose as quota X, the number of dots in each of the upper m rows, since in the lowest row there are fewer than X dots, this condition will be *sufficient* to ensure that not more than m candidates can be elected.

Next suppose that we take away one dot from each of the upper m rows, and place the dots so obtained in the lowest row. The lowest row will now contain $(X - t^* - m) \geq (X - 1)$ dots; that is, in the new graph the lowest row will contain at least as many dots as there now are in each of the upper m rows. It follows that the number of dots in each of the upper m rows of the original graph, must give the *smallest* number of votes which can prevent the election of as many as $(m + 1)$ candidates.

Obviously the number of dots in each of the upper m rows is an integer.

Thus a dot graph constructed in the manner described to contain a total of e dots, will have X dots in each of the upper m rows, such that:

(i) X is an integer,

(ii) chosen as quota, X will prevent the election of as many as $(m+1)$ candidates, and

(iii) X is the smallest number which, chosen as quota, will do so.

Thus X defines a size of quota which has precisely the set of properties which is necessary and sufficient to define the Droop quota, and $X = Q_d$.

The number of dots in the lowest row can be expressed as $(e - mX) = (e - mQ_d) = (Q_d - t^*)$, $1 \leq t^* \leq (m+1)$. Earlier the algebra had shown $(e - mQ_d) = (Q_d - t)$, $1 \leq t \leq (m+1)$. Hence $t^* = t$.

It can be shown that a graph of this kind can be constructed if and only if the number of dots in each of the upper m rows is at least equal to the number of crosses in the lowest row; that is, if and only if $e \geq mt$, the same condition as was given by the algebraic treatment.

Granted such a graph can be constructed we can read off:

(i) the total number of voters, e, as the total number of dots,

(ii) the number of seats in the constituency, m, as the number of rows, each containing the same number of dots, above the lowest row,

(iii) the size of the Droop quota, Q_d, as the number of dots in each of these m upper rows, and

(iv) t, as the number of crosses in the lowest row.

Thus one way of setting out the data of the problem in PR is to make a statement such as, say, $e = 100$, $m = 3$, and using algebra to derive from this $Q_d = 26$, $t = 4$. An alternative way is to provide a dot graph. We need to check that each row above the lowest row has the same number of dots, and the number of such rows gives m. And we need to check also that the number of crosses in the lowest row lies between 1 and $(m + 1)$. When these two tests are satisfied, we can read off from the graph the sizes of e, m, Q_d, and t. To each valid statement that can be made in terms of algebra, a corresponding statement of the same import can be got by interpretation of the dot graph. The algebraic treatment and the graphical treatment are logically equivalent to each other. Hence

Proposition 3. A dot graph consisting of m rows, each with the same number of dots, and a further row containing between 1 and (m + 1) dots fewer than any of the preceding rows, and possibly empty, is a Droop representation from which the values e, m, and the corresponding values of Q_d and t can be read directly.

Graphical interpretation of the defining property of the Droop quota. Its defining property is that this quota shall be the minimum number of votes which is sufficient to ensure that $(m+1)Q_d > e$. This yields the formula $Q_d = (e+t)/(m+1)$, $1 \le t \le m+1$.

The Droop representation we have described, has a Droop quota of dots in each of the upper m rows. Now take away one dot from each of these upper m rows and place the m dots got in this way, in the lowest row. Each of the upper m rows now contains $(Q_d - 1)$ dots, and the lowest row contains $(Q_d - t + m)$ dots, $1 \le t \le m + 1$, and so contains at least $(Q_d - 1)$ dots, that is contains at least as many dots as each of the upper m rows. A quota of the size $(Q_d - 1)$, therefore, would be held by at least $(m + 1)$ candidates, and so would infringe the definition of the Droop quota. So also would any quota got by taking away two dots from each of the upper m rows and placing them in the lowest row, and so on.

Now we may envisage a dot graph in which each of the upper m rows has the same number of dots, this being more than one Q_d; and if, in a graph of this kind, we were to reduce by one the number of dots in each of the upper m rows, we might find that the graph obtained was still able to satisfy the basic condition in the definition of the quota. But continuing in this way to reduce the number of dots in each of the upper m rows, a point would be reached at which any further reduction in the number of dots per row, would lead to infringement of the condition.

If we choose, therefore, we may regard the Droop representation as being got by finding the particular graph, with the same number of dots in each of the upper m rows, which gives rise to the minimum number of dots per row which satisfies the condition stipulated in defining the quota.

Proposition 4 (on the dynamics of the Droop representation), For $e > mt$, m given, if the number of voters is increased by t or more, the size of the Droop quota is increased; and if the number of voters is reduced by $(m + 2 - t)$ or more, the size of the Droop quota is diminished.

For any of the $(m + 1)$ values given by an increase in e of $(t - 1)$ or less, or given by a reduction in e of $(m + 1 - t)$ or less, the size of the Droop quota will remain unaltered. Since $e > mt$, a Droop representation will exist for all the values of e under consideration, and the lowest row of each such representation must contain at least t crosses, $1 \le t \le m + 1$.

If the size of the Droop quota remained unaltered, increasing e by t or more would give rise to at least as many dots in the lowest row as in each of the upper rows, and no crosses; and a figure of this description would not be a Droop representation. But a Droop representation exists; and the only way in which it can be arrived at is by the transference of at least m dots from the lowest row to increase the number of dots in each of the *m* upper rows by at least one. That is, the size of the Droop quota will increase.

If the size of the Droop quota remained unaltered, reducing *e* by at least $(m + 2 - t)$ would give rise to more than $(m + 1)$ crosses in the lowest row and the resulting figure would not be a Droop representation. But for the new value of *e*, a Droop representation exists. It can be got by the transference of at least one dot from each of the m upper rows to the lowest row. Thus for the new lower value of *e*, the size of the new Droop quota will be lower than the initial size by at least one.

An increase in the value of *e* by 0 or 1 or 2 or ... or $(t - 1)$, will add the same number of dots to the lowest row, but will leave at least one cross in that row and the size of the Droop quota will remain unaltered. Of course for $t = 1$ an increase in *e* by one or more will give rise to an increase in the size of the Droop quota.

A reduction in *e* by 1 or 2 or ... or $(m + 1 - t)$ will still leave fewer than $(m + 1)$ crosses in the lowest row, and will leave the size of the Droop quota unaltered.

In all $(m + 1)$ values of *e*, including the original value, give the same size of Droop quota.

	$e =$	67	68	69	70	71	72	73
$m=3$	$Q_d =$	16	17	17	17	17	18	18
	$t =$	1	4	3	2	1	4	3

Figure 3.7

The table of fig. 3.7 takes $m = 3$. Supposing the initial value of *e* to be 70, giving $t = 2$, then an increase in *e* by one $(= t - 1)$ leaves the size of the quota unaltered, and an increase in *e* by two $(= t)$ gives an increase in the size of the quota by one. A reduction in *e* by $(m + 2 - t)$, that is by three, from 70 to 67, gives a fall in the size of the quota by one. For the $(m + 1) = 4$ successive values of *e* lying between 68 and 71, the quota has the same size.

The lowest row in the table (Fig. 3.7) shows that as *e* increases, an increase in the size of the Droop quota corresponds to the occurrence of $(m + 1) = 4$ crosses in the lowest row of the dot graph. Further increases in *e* by one unit at a time, gives $m = 3$ crosses in the lowest row of the dot graph; then two crosses; then one cross; then four crosses again, and so on.

The Droop paradox (non-monotonicities). In 1881 'it was found that, with the Hare quota in use, an increase in the membership of the House of Representatives from 299 to 300, would have given the state of Alabama 7 members in place of 8, and this was termed 'the Hare paradox''.[13] The Droop quota has the similar

13 Cf. Clarence G Hoag and H Hallett, Jr., *Proportional Representation*, p. 417, and Robert A Newland, 'Hare Quota Dead but still Running', *Representation*

feature that for some configurations, as A, the number of supporters of the party A, increases from say, A_1 to A_2, while the number of B-supporters remains constant, instead of increasing, the number of Droop quotas got by A will diminish. A may shed a member, as the result of an increase in its number of voters.

Using the subscript "1" to denote the initial state of affairs, and "2" to denote the final state of affairs, take $A_1 = 4000$, $B_1 = 1998$, $m = 5$; then $e_1 = 5998$, $Q_{d1} = 1000$, $t_1 = 2$. A gets 4 complete quotas *plus* 0 additional votes.

For $A_2 = 4003$, $B_2 = B_1 = 1998$, $m = 5$; and $e_2 = 6001$, $Q_{d2} = 1001$, $t_2 = 5$. After the increase in its numbers, A now gets three complete Droop quotas *plus* 1000 votes, but these 1000 votes do not amount to a complete quota. Initially A had filled four seats and finally it will fill three.

As another instance take $A_1 = 120$, $B_1 = 79$, $m = 9$; giving $Q_{d1} = 20$, $t_1 = 1$, and $A_1 = 6Q_{d1}$ plus zero votes.

Then let $A_2 = 131$, $B_2 = B_1 = 79$, $m = 9$; giving $Q_{d2} = 22$, $t_2 = 0$, and $A_2 = 5Q_{d2}$ plus 19 votes. In spite of the increase in its numbers by 9%, while B remains constant, the party gets fewer complete Droop quotas than initially.

Obviously a condition necessary for the occurrence of the Droop paradox is that Q_{d2} must be larger than Q_{d1}; that s, the size of the quota must be larger for $e = e_2$ than for $e = e_1$. Or, if we represent the first state of affairs by a dot graph, and the second state of affairs by another dot graph, the second graph must have at least one dot more in each of the upper m rows than the second graph. It can be shown that a necessary condition for their occurrence is that $m \geq 5$, and that the non-monotonicities conform to a geometrical pattern. But adoption of the Games solution for the allocation of seats to the parties, as is done by Carroll (see below), avoids this difficulty of non-monotonicities, and meantime we won't try to follow up the paradox further.

3.5. Walter Baily and the Number of Voters Unrepresented

Droop thus suggested a particular size of quota which, for use in elections, would have some very valuable properties. In particular it would prevent more candidates being elected than there were seats available to fill. Walter Baily, Droop's brother-in-law, found that under a two-party system the Droop quota possessed the highly interesting property that it minimised the number of 'wasted votes' - votes which had no influence on the election result. He gave an account of a recent School Board election to fill seven seats, in which almost half the total number of votes cast had been 'wasted' and had been without influence on the election result. Baily adopted the criterion that an optimal electoral procedure would minimise the

---

(Journal of the Electoral Reform Society), Vol. 20, No. 78, Jan. 1980. [The definitive discussion of this controversy is M Balinski and H P Young, *Fair Representation: meeting the ideal of one man, one vote* (New Haven, Yale University Press, 1982)]

number of votes wasted. Carroll's work takes some of its positions straight from Baily and indeed without knowing Baily's pamphlets, it would be exceedingly difficult to understand how Carroll takes some of the issues that he does.

Walter Baily was the author of two pamphlets, *A Scheme for Proportional Representation* and *Proportional Representation in Large Constituencies*. By citation of parallel passages and by comparisons of language and argument, it is easy to show that at a number of points Carroll is under obligation to the later of these two pamphlets. In particular he adopts Baily's conception of 'wasted votes', and from first to last this becomes a main strand in Carroll's argument.

Baily attributes the lack of correspondence between the number of votes cast for the parties and the results of an election to vote 'wastage'. He believes (or seems to, for he does not express it quite in these terms) the vote wastage is the necessary and sufficient condition for failure to secure proportional representation.

He starts by considering a recent School Board election, held under a procedure of cumulative voting, and computes that about half the votes cast had been wasted in one way or another. In a system of preferential voting (such as we are familiar with today in the single transferable vote):

> The number of votes to be retained for a candidate must be enough to make his election certain, whatever combination be made of the other votes given in the election; *the smallest number that will suffice for this is the true Quota: all votes retained beyond this number are wasted.* (see below, p. 181, my italics).

And as we know from the foregoing theory, this results in the choice of the Droop quota.

Even with the Droop quota in use there is vote wastage, but its scope is now at a minimum and its extent is subject to a calculable limit. To illustrate by an example in which, with preferential voting in use, $e = 100$, $m = 2$ and Droop quota = 34. If two candidates are elected, using up 34 votes each, and over and above will remain 32 votes which have played no part and are 'wasted votes'.

Baily himself was opposed to the use of any procedure which involved preferential voting in elections, giving a reason which was to be parodied by Carroll in his passage about 'Hodge, fresh from the plough' (*St James's Gazette*, 4 July 1884)

> I am disposed to think, notwithstanding the sanguine views of Mr Hare, and several of his followers, that there would be a good deal of inconvenience in using [preferential voting] for large constituencies where the voters would be counted by thousands, and would be but too often imperfectly educated.[14]

---

14 Walter Baily, *Proportional Representation in large constituencies*, p. 9.

The criterion of a suitable election procedure that Baily adopts, is that

> *A sound system of election must provide, as much as may be, against....waste*[15].

In the second half of his pamphlet Baily proposes a list system of PR, in which a candidate is required, for election, to get a Droop quota of votes. Votes are to be transferred, much in the manner of the single transferable vote, among the lists (schedules of preferences) of the various parties, in accordance with a pre-established sequence; and Baily's arithmetical illustration involves the seven different parties which we may term A, B, ...,G.

Since the qualification for election is a quota of votes, when all the available seats have been filled, a total of $(Q_d - t)$ votes will have been taken up by no candidate; and Baily refers to this as 'the number of votes unrepresented'. Proceeding beyond this, if ,say, the quota is 1200 and some particular candidate has been elected by getting 100 votes from the party B, 700 votes from D and 400 votes from G, in Baily's term this candidate is said to *represent* 100 B-votes, 700 D-votes and 400 G-votes. Doing this, a final table can then be constructed to show 'the total number of votes represented' in any party: this will be the total number of votes taken, at one stage or another, from the list of candidates for this party, in the election of the candidates who in fact get the seats. But a figure of this kind is in the nature of a book-keeping item and has little counterpart in reality.

The terms 'the number of Electors represented' and 'the number of Electors unrepresented' are in themselves enough to show that Carroll was influenced by this part of Baily's scheme. But Carroll was working within a two-party system, and the strongest objections which can be brought against Baily's 'number of votes represented' do not apply against Carroll's 'number of Electors represented'. He had given Carroll a useful start in his theorizing, but the main development which Baily himself went on to give, would seem to be specious. It is also highly complicated and would not be worth attempting to follow up further.

3.6. J G Marshall and the Two-Person Zero-Sum Game

> *The hardest problem of a young mathematician is to find a problem. The right question, well asked, is more than half the battle, and often the only part that requires inspiration ... It often happens that all the thrill of creation and insight is concentrated in the question.*
>
> *... After the question is formulated, the mathematician does not proceed ... like a scientific Sherlock Holmes. A mathematician is not a deductive machine, but a human being. New mathematics comes to him not by pure thought and deduction, but by sweat, experiment, induction,*

15 *Op. Cit.* see reprint below, p. 180.

> *and, if he is lucky, by inspiration. Of course a mathematical experiment does not involve wires, tubes and bubbling liquids; it consists rather of a detailed examination of some particular cases or analogues of the desired result ... On the basis of such experiments the mathematician jumps inductively to hold conclusions. It may be a difficult task to prove them, but often the purely deductive arrangement of the work serves more to communicate facts than to establish them.*[16]

Life. J G Marshall (1802--73) was the son of a wealthy flax-spinner who had been MP for Leeds. Marshall attended school in Edinburgh and afterwards attended classes in the university, but did not take a degree. Later he too entered the flax-spinning business and, like his father before him, became MP for Leeds.

He felt attracted to the sciences, particularly to Geology, and a powerful scientific bent led him to devise improvements in flax-spinning machinery, but he did not take out patents for these improvements, preferring that the benefits should be available to people generally, without restriction. He was a Liberal and keenly interested in reform movements, with his main passion reserved for electoral reform. His pamphlet *Minorities and Majorities: their Relative Rights* was followed by two others of a more general nature in 1861 and 1869, but it is only with *Minorities and Majorities* that we will be concerned.

The 2-person game derives from Marshall. Before his trip to London late in May 1884, Carroll's work on PR showed little promise: but within days of his return, he had adopted various ideas from a pamphlet by Walter Baily, and his thinking got off the ground. The value of Carroll's booklet, however, lies in its use of the two-person zero-sum game: this gives shape and depth to Carroll's reasoning, and holds the argument together; and it is hard to conceive of the booklet existing at all without the concept of the two-person game. At an earlier date the present writer had taken the view that this portion of Game Theory was of Carroll's own invention and had praised his work correspondingly.[17] But the evidence which we now present would seem to leave no doubt that the view of the two-person game on which Carroll's booklet is based, had been the work of J G Marshall. The present section will largely take the form of showing that this is so.

Our own book has been an exposition of Carroll's theories, and, if we are right in the above opinion, it will at the same time have been an exposition of Marshall's theory. This allows the discussion we give of Marshall's work to be briefer. A reprint of the portions of his pamphlet relating to the two-person zero-sum game is given belowm [pp. 175--179].

---

16 From Paul R Halmos, 'Innovation in Mathematics', reprinted in Morris Kline ed., *Mathematics in the Modern World: readings from the Scientific American*, (San Francisco, W H Freeman, 1968), p. 9.

17 [See Part 2, p. 47].

The external evidence. In chapter VII of his *Autobiography* John Stuart Mill had mentioned that his *Thoughts on Parliamentary Reform* (1859) had made:

> a claim of representation for minorities; not, however at that time going beyond the cumulative vote proposed by Mr. Garth Marshall.[18]

This reference by Mill in itself would probably have led Carroll to seek out Marshall's work, and besides this he would have come across other references to the pamphlet. In particular he would probably have read the able article by G Shaw Lefevre in the common-room copy of *The Contemporary Review*, May 1884. On the whole the external evidence suggests that it is likely that Carroll would have seen Marshall's pamphlet.

The internal evidence: two principles. The following passage from Marshall's pamphlet relates to the procedure which gives the elector one vote. It states two principles which are central to Marshall's theory, and which later were to become key principles for Carroll. It is the solitary passage in which Marshall gives anything approaching a connected account of his theory. As we will explain later, it is supplemented by a number of arithmetical examples.

> Adopting what is called the Single Vote, each elector can vote 'only' for one representative. The minority, whatever it might be, would then have the power, by a proper distribution of their votes, of securing a fair share of the representation. If we suppose the counting to be divided between two parties, the smaller being, for example, one-third of the whole, if that minority *were to restrict their list of candidates* to one-third of the number to be elected, and *were to distribute their votes equally among them*, [it] might thus secure the return of their list of candidates. *If, again, we suppose the country to be divided into an almost indefinite number of parties*, this mode of voting would *in like manner* give a proportionate share of representation to each. For instance, any party or opinion in this country which could number 1/656 part of the electors amongst its supporters, might by combination return one Member and have one voice in the Legislature.[19]

In the course of this passage Marshall formulates the (Droop) quota in the fractional form which later was to be adopted by Carroll.

The other principle which makes its appearance in this passage is even more far-reaching. To maximize the number of seats that they fill, the supporters

---

18 Cf. *Autobiography of John Stuart Mill published from the original manuscript in the Columbia University Library*, with a preface by John Jacob Coss (New York: Columbia University Press 1924), p. 180.

19 J G Marshall, *Minorities and Majorities*, p. 12.

of a party may 'restrict their list of candidates' and 'distribute their votes equally amongst' the candidates chosen. This is the proposition, which we have dubbed 'J G Marshall's Theorem', and does not call for further discussion.

We will now try to indicate the main feature of Marshall's theory for purposes of comparison with the structure of the argument of Carroll's booklet.

Characteristics of Marshall's work. In Marshall's hands the problem of PR is narrowed down to a single issue - the contest between two opposing parties, in a constituency which, he recommends, would have at least three members. Each party seeks to maximize its number of seats. The number of seats filled by a party will, of course, depend on the number of its supporters; and, besides this, will depend on the number of party candidates who stand, and on the distribution of the supporters' votes amongst these candidates.

With a given number of candidates in the field, Marshall stated clearly that an optimum distribution would spread the votes for a party, as evenly as possible over an appropriate number of candidates. It is obvious that if a large enough number of candidates should stand, a party making an even distribution of votes will fail to fill any seat; and Marshall's arithmetical examples stress the need for *the number of candidates* being in some sense an 'optimum'. The problem then admits the simple enunciation: given the number of party supporters and given that there will be an even distribution of votes among the candidates, what is the 'optimum' number of candidates for the party?

The breakthrough originally for Marshall and later for Carroll, may have been in visualising the problem in these simple terms; and the student might do well to concentrate on this as a bird's-eye view of the problem.

Marshall had been able to devise this conceptual scheme or model for the two-party contest, but he encounters great difficulty in communicating his vision of the two-party game to the reader. At no stage in the discussion does he employ any mathematical symbol. The pamphlet endeavours to express propositions of mathematical purport, without using the only language adapted to the task, and it is never easy to grasp Marshall's meaning. The reader must them turn to the set of arithmetical examples he provides as Appendix C to the pamphlet[20]. This considers the operation of two procedures, the 'single vote', giving the elector one vote, and the 'cumulative vote' which, in the m-member constituency ($m \geq 2$) gives the elector m votes which he may divide among the candidates in any manner he chooses. All the answers Marshall gives to the problems he poses are correct and his solutions throw light on the model in terms of which he views the problem: but an arithmetical treatment must be distressingly dull.

A simplification and an extension. Marshall had shown the damage which a party might sustain from having too many candidates in the field, and from an uneven spread of votes among the candidates. Now, writing after the experience of the

20 [Reprinted below, p. 175--179].

Birmingham Caucus, Carroll was able to make certain simplifications on Marshall's scheme of things. He supposed that:

(a) a party or its managers (he makes no attempt to specify the mechanism involved) has complete knowledge of the number of its own supporters and that of its rival

(b) the party is able to put up some definite number of candidates in the election, and

(c) with a view to an even spread of votes among the candidates, the party can arrange which of its supporters shall vote for which candidates.

With the Birmingham Caucus continuing to operate, most readers could be expected to accept these assumptions as being reasonable and realistic. They enabled Carroll to formulate a version of the two-party game which incorporated the essentials of Marshall's discovery, and which at the same time is simple enough to be expressible in terms of linear algebra. It also has the inestimable virtue that, apart from ties in the voting, it leads in every instance to a determinate solution.

Once the theory of the single vote had been formulated algebraically (we may surmise) Carroll saw an extension of its scope that would present little difficulty: the theory could be got to cover procedures which, instead of allowing the elector only one vote, would allow him v votes, $1 \leq v \leq m$, of which not more than one could be given to any particular candidate. The theory covering this entire family of procedures was expressible in the exceedingly brief terms of four linear algebraic inequalities. These provided the heart of Carroll's whole theory of PR, and are summarised in the fourfold table (see p. 159). The cumulative vote with which Marshall was concerned, gives the elector m votes and places no restriction on whether he gives them to m or fewer candidates. It falls outside the family of procedures we have just defined: but when we take into account Carroll's assumptions (a) - (c) its operation can be seen to differ only minutely from that of the single vote. At Carroll's hands it calls for only the briefest mention (page 165), and Marshall himself had thought that, in purely arithmetic terms, the two procedures worked out identically (page 182).

An example from Marshall. To assist the student in tackling the assortment of examples in Appendix C of Marshall's pamphlet, we will work out the first of his examples. If, as in the present text, we denote the two parties by A and B (though in Marshall's usage these letters are the names of candidates) his first example takes $m = 2$, A = 200, B = 100. Under the cumulative vote there are 400 A-votes and 200 B-votes. Without discussing how this comes about, Marshall assumes that there will be two candidates for the party A and one for the party B. The theory given in our own text shows an optimum strategy for A to be A{200, 200} and for

B to be B{200, 0}. A is certain to fill one seat and will tie with B for the other. Marshall's conclusion is that with the cumulative vote in use, 'If the minority *exceed* 1/3, they have half the Representation, if they have less than 1/3 they have no Representation' which is a fairly comprehensive answer.

With the single vote in use, on the theory given in the present book, the optimum strategies for the parties are A{100, 100} and B{100, 0} respectively, giving the same outcome as for the cumulative vote.

The time span. Carroll's earlier work on PR left him, late in May 1884, starting virtually from scratch, and his article in *The St. James's Gazette* of 10 October presented a shortened, but fairly comprehensive, version of the theory of the booklet. Had Carroll done so, it should indeed have been a work of pure genius to have completed a task of this kind in the space of 140 days. Within that time Carroll had indeed done much. But the vital element, the two-person game of the booklet, had been adapted from the pamphlet of J G Marshall. Carroll's failure to acknowledge the work of Marshall which had in effect provided the valuable element in his booklet, was ungenerous and deplorable.

The James Garth Marshall Theorem. As efficient a way as any for a party to fill s seats ($1 \leq s \leq m$), is for it to put up exactly s candidates and divide its votes among them as evenly as possible.

There may be other strategies by which the party could fill s seats; but if any such strategy exists, the party could also fill s seats by using the strategy described. We will refer to a strategy of this kind as being an optimum strategy of the party to fill s seats.

Each party will have m optimum strategies of this kind. We need not introduce into the theory any discussion of an optimum strategy to fill no seat. This will be taken care of automatically by the other party being able to fill all m seats.

Proof of theorem. consider the case in which the m-member constituency is contested by p parties, $p = 2$. The elector is allowed a single vote, we assume, and the m seats available are awarded to the m candidates who have the highest numbers of votes. Let us suppose that the party A, under some given uneven distribution of votes over its candidates is able to fill s seats ($1 \leq s \leq m$). If so, the A-candidate with the sth-highest vote among its candidates, is elected.

Now consider any more even distribution of its votes among its candidates. The A-candidate with the sth-highest vote will have at least as many votes as had the A-candidate with the sth highest vote in the original distribution. The votes of the B-candidates, C-candidates, and so on, will remain as they were in the original distribution. Hence the A-candidate with the sth-highest vote in the new distribution will again fall into the group of m candidates with the highest numbers of marks, and will be elected. But if so, so also will be the A-candidates with the $(s - 1)^{th}$ highest mark, the $(s - 2)^{th}$ highest mark, and so on, in the new distribution. And if

A is able to return s candidates with the less even distribution of its votes, it must also be able to return s candidates with its more even distribution.

This theorem is directly applicable to the two-party zero-sum game, and equally applicable it will be found, to the p-party zero-sum game.

3.7. Demand Curves, Maximin, and the d'Hondt Scheme of PR

The demand curve A+B is got by lateral addition of A and B. One of the fundamental theorems of Economics shows that the curve of total demand for a commodity in a market, is got by lateral addition of the demand curve for the commodity of the individuals in the market.[21]

Applied in Politics. In the case of political parties we will consider the demand originating from the parties for the seats available in the election in the constituency; and the price will be in terms of the number of votes per seat that a party is willing to pay. The total demand of the parties for seats, at any given price per seat, will be the sum of the demands coming from the parties at that price. In the present instance our supposition will be that the constituency is contested by only the two parties A and B. At any given price per seat, therefore, the total demand for seats will be the number of seats demanded at that price by A, plus the number demanded by B.

Also our assumption will be that the two parties are *price-takers*. A price is named, so many votes per seat, and each party states the number of seats it is willing to take up at that price. This will depend on the number of votes that the party has. A party has a certain number of supporters who vote for it. If A has 100 supporters, with $v = 1$ (as the theory will suppose, to begin with) A will have 100 votes. A has nothing on which to spend these 100 votes, except in acquiring seats. If the price at which seats are offered is, say 50 votes per seat, A will signify that it is willing to take up two seats at that price. If the price were 40 votes per seat, A would be willing to take up $100/40 = 2.5$ seats - but is precluded from doing this by the fact that seats are indivisible and can be transferred only as an integral number at a time A's demand for seats will remain at the total of 2 seats, until the price is reduced to 33 seats per seat, when A's demand for seats will extend to 3. A would have demanded 3 seats when price fell to 33.3 votes per seat, but for the fact that, like seats, votes too are indivisible and can be offered and accepted only as integral numbers at a time; this is possible at 30 votes per seat but 33.3 votes is not a possible price.

---

21 Cf. Léon Walras, *Elements of Pure Economics* (translated by William Jaffé, London, Allen and Unwin, 1954) pp. 92--106. Walras' work appeared originally in 1874.

A party's curve of optimum strategies. A party, we have shown, will do at least as well in filling any given number of seats, by employing the 'optimum' strategy appropriate to that number of seats, as by any other.

Example. In a constituency with 6 seats, the party A has 171 supporters and B 80 supporters. Each elector is allowed a single vote. How many seats will be filled by each of the parties?

If A puts up one candidate, it will give him 171 votes. With two candidates, it could give to each 171/2 = 85.5 votes, but for the fact that a candidate must receive a whole number of votes: and the party may give 86 votes to the one candidate and 85 to the other. To let the argument proceed unencumbered, we will meantime neglect the fractional part in the number of votes, and take it that in a case like this each candidate is given 85 votes. (The adjustment to the argument needed to take into account that in fact one candidate will get 86 votes and the other 85, is gone into later).

With three candidates the party A will give to each candidate 171/3 = 57 votes, to each of four candidates 171/4 = 42.75, say 42 votes, to each of five candidates 171/5 = 34.2, say 34 votes, and to each of six candidates 171/6 = 28.5, say 28 votes.

We plot this information to get A's optimum-strategy curve in Fig. 3.8, in which along the horizontal axis we take the number of candidates, and along the vertical axis we take the number of votes per candidate. We join the points plotted by a sequence of vertical and horizontal straight lines, shown as broken lines, and we place alongside this step curve the name of party A. Each step on the curve has the same width, viz. one candidate as measured on the horizontal axis, and, as we move down the curve, each step becomes shallower. The curve is to be considered real only at the point at the edge of each step, the joining lincs bcing addcd merely to assist the eye.

Consider next the party B. If it puts up 1 candidate it can give him 80 votes, if it puts up two candidates it can give each 40 votes, to three candidates it can give each 26.7, say 26 votes, to four candidates it can give each 16 votes, and to six candidates it can give each 13.3, say 13 votes.

We likewise plot these points in Fig. 3.8 to get B, the optimum-strategy curve for this party; and we join up the points by a sequence of vertical and horizontal straight lines, shown as dotted lines to differentiate it from the curve A.

At the same height on any two such curves A, B, the one which is further from the vertical axis tends to have the shorter vertical steps.

Next in the diagram, we construct the curve A + B which is got by lateral addition of the curves A and B; that is, for any given ordinate, the abscissa of A + B is equal to the sum of the abscissae of the curves A and B.

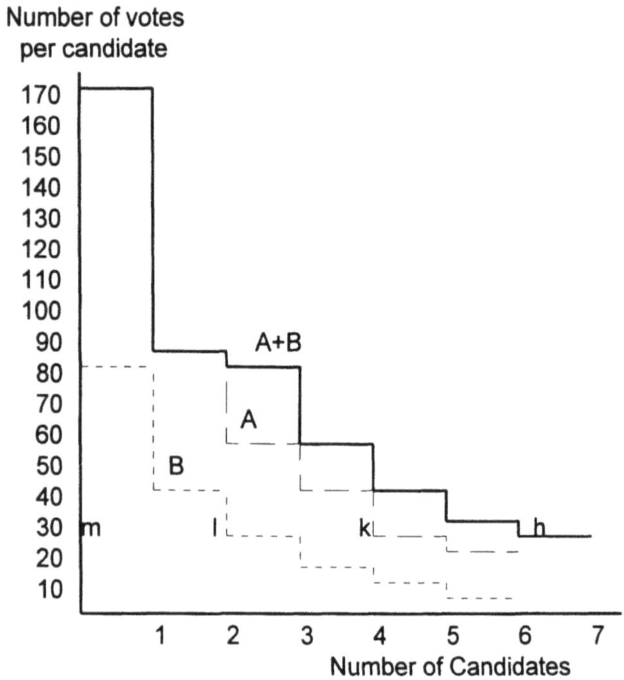

Figure 3.8

Since each of the curves A and B is real at, and only at, integral values on the horizontal axis, so also the curve A + B is real at, and only at, such values, i.e. for the values 1, 2, .. candidates. Also A + B will have real values at, and only at, those ordinates at which A or B has a real point. From the way in which it is formed[22] A + B will be a step curve consisting of vertical and horizontal segments passing through real points, and we show these segments as continuous lines. If A and B happen each to have a step at the same height, A + B will have a single step at this height containing two real points, one at the middle of the step and one at the edge. Apart from the case in which the steps coincide in height, A + B, over any given range of the vertical axis, will have as many steps as A and B jointly. Although as we move down A or B the steps on the curve get monotonically shallower, as we move down A + B there will as a rule be irregularities in the rate at which its steps become shallower, since the steps on the constituent curves will sometimes occur closer together and sometimes further apart.

In the example six seats are available: to show the allocation between the two parties, through the point six on the horizontal axis, draw a vertical to intersect

---

22 The reader may consult, on the whole of this technique, Léon Walras, *Elements of Pure Economics*, pp. 92--106.

A + B in the point *h* say. From the way in which A + B has been constructed, *h* must be a real point and must correspond to an integral number of votes per candidate. Through *h* draw a horizontal to intersect A, B and the vertical axis in *k*, *l* and *m* respectively. The one party will fill *mk* seats and the other *ml* seats, this giving in Fig. 3.8, four seats to A and two to B.

We can satisfy ourselves arithmetically that, with the volume of support it has, this is the best result that either party can achieve against intelligent strategy on the part of its rival. If A puts up four candidates, it can do at least as well by dividing the available votes among them as by any other strategy, giving each 42 votes. If B should then put up one candidate, it would of course fill only one seat. If B puts up two candidates, it can give each 40 votes, and fill two seats. If, while A puts up four candidates, B put up three candidates, each with 26 votes, it would still fill two seats: but if A then put up five candidates, each with 34 votes, B would fill only one seat. In fact, confronted with the strategies open to A, B cannot do better than put up two candidates, so filling two seats.

Next, if A put up fewer than four candidates, it would fill fewer than four seats. If A put up five candidates, giving each 34 votes, against B's retort of putting up two candidates, each with 40 votes, A would still fill only four seats, and so on. In fact, confronted with the appropriate counter-response on the part of B, no strategy exists whereby A can improve on its position as given by the geometrical construction.

Taking into account fractions in the number of votes per candidate: the interpretation of the diagram. The party A, dividing 171 votes between two candidates, could give each 85.5 votes, but for the fact that each candidate must have an integral number of votes. Hitherto we have chosen to regard a party in this position as giving 85 votes to each candidate although, to use up all its votes, it would in fact give 86 to one and 85 to the other. Putting up four candidates, 171/4 = 42.75; and dividing the votes as evenly as possible among the candidates, the party would give 43 votes to each of three candidates and 42 to the remaining candidate, and similarly in other instances.

Disregarding fractions in the number of votes per candidate, as we have done hitherto, gives an approximation to the actual state of affairs, and enables an optimum-strategy curve to be drawn, consisting of a series of plateaus, each standing at the height of an integral number of votes.

A distribution of votes among the candidates such that no candidate receives more than a single vote more than another is an optimum distribution. To take the fractional part of the average vote into account, therefore, we need only interpret the optimum strategy curve in this way: the height of the curve on any plateau shows the number of votes that at least one of the candidates receives, but some, though not all of the candidates concerned, may get one vote more than this: e.g. in Fig. 3.8 for the plateau shown at height 85, one candidate gets 85 votes and the other 86; and for the plateau at height 42, one candidate gets 42 votes and the three others 43 votes each.

The plateaus are an approximation to the actual state of affairs correct to within one vote per candidate, and exactly correct in each case for at least one of the candidates concerned.

Lewis Carroll's Theorem. If h lies at the right-hand extremity of a step on A+B, the geometrical construction shows the allocation of seats among the parties when they choose in accordance with the maximin criterion.

The (rough) economic proof of Carroll's Theorem. Suppose we take an election contested by three parties, say, in which there are four seats available and each elector is allowed one vote. Each party has a certain number of supporters and can do what it likes with their votes, dividing them among its candidates as it chooses.

The four seats available go to the highest bidders: the price of a seat becomes so many votes per candidate, just as in the case of a commodity it would be so many shillings per lb. The 'money' or stock of votes that a party has is of no use to it except for the purchase of seats, and it will be willing to spend all the votes that it has, spreading them over the candidates that it puts up. A party's optimum-strategy curve is equivalent to a 'demand curve' for seats. The party is spending all its votes on 1 or 2 or ... candidates, and, when we disregard fractions in the numbers of votes, this demand curve will have the form of a step curve.

Incidentally this is also the form of curve envisaged by Cournot and Walras for the individual's demand curve for a commodity. Thus Cournot: 'In a certain household the same quantity of firewood will possibly be used whether wood costs 10 francs or 15 francs the stere, and the consumption may suddenly be diminished if the price of the stere rises above the latter figure'[23]. But whereas it was for empirical reasons that these authors envisaged the individual's demand curve as being a step curve, in the political problem a party's demand curve for seats has this shape necessarily, on account of indivisibilities: the bidding must be for an integral number of seats and a candidate must be given an integral number of votes. Apart from the requirement that each candidate be given an integral number of votes, the real points on a party's demand curve for seats would lie on a rectangular hyperbola.

The constituency with a definite number of seats, in this instance four, corresponds to the market in which a fixed stock of the commodity is available: the supply curve is a vertical straight line through the corresponding point of the horizontal axis[24].

---

23 *Researches into the Mathematical Principles of the Theory of Wealth*, p. 50 in Nathaniel T Bacon's translation (New York, Macmillan, 1897) of a work that had appeared in 1838; cf. also Léon Walras, *op. cit.*, p. 95f.

24 The classical statement of the economic theory is Philip H Wicksteed, 'The Scope and Method of Political Economy', *Economic Journal*, Vol. 24, 1914, reprinted in *The Common Sense of Political Economy*, ed. Lionel Robbins, (London, Routledge, 1933), Vol. 2, pp. 772--96.

If, in this 4-seat constituency, the elector is allowed say, three votes to be given to different candidates, each of the optimum-strategy curves in the diagram, i.e. each party's demand curve for seats, has a topmost plateau consisting of three real points. If a party has say, 73 supporters, it can pay 73 votes per candidate for each of the 1st, 2nd and 3rd seats[25]. After that it can divide up 73 x 3 = 219 votes among whatever number of candidates it puts up; beginning with the 3rd candidate, the curve approximates to a rectangular hyperbole. The theory again works out after the pattern of demand and supply in Economics.

The available seats are allotted to the highest bidders, but the allotment among the parties must be made simultaneously, not by one party buying a seat now and another buying a seat a little later. The election mechanism is equivalent to a market mechanism in which the price per seat is 'cried'[26] and each party indicates the number of seats that it bids for at that price. When a price, so many votes per seat is found, that just clears the market of those available, the allocation of seats among the parties becomes definite.

Alternatively the economic problem of the market and with it the political problem of the allocation of the seats in a constituency, can be envisaged as a process of contract and recontract, until a stable set of contracts is arrived at[27]

Carroll's theorem allocates seats in accordance with the d'Hondt quota. The set of rules proposed in 1882 by Victor d'Hondt[28] a professor in the University of

---

25 Alice was able to get two eggs cheaper than one, but had reason for buying only one.

> 'I should like to buy an egg, please' she said timidly. 'How do you sell them?'
> 'Fivepence farthing for one - twopence for two', the Sheep replied.
> 'Then two are cheaper than one?' Alice said in a surprised tone, taking out her purse.
> 'Only you *must* eat them both, if you buy two', said the Sheep.
> 'Then I'll have *one*, please', said Alice as she put the money down on the counter. For she thought to herself 'They mightn't be at all nice, you know'.

(*Through the Looking-glass*, chapter 5)

In the market in seats the party with most support gets three seats for the price of one, and its allocation of seats is disproportionately greater than the volume of its support.

26 Cf. Léon Walras, *op. cit.*, pp. 83--6.

27 Cf. F Y Edgeworth, *Mathematical Psychics* (London, Kegan Paul, 1881), especially pp. 16--9, George J Stigler, *Production and Distribution Theories* (New York, Macmillan, 1941), pp. 244--5, and T W Hutchison, *A Review of Economic Doctrines 1870-1929* (Oxford, Clarendon Press, 1953), pp. 113 and 206.

28 Cf. C G Hoag and G H Hallet, *Proportional Representation*, pp. 418--24, who give valuable references to the literature. Cf. also d'Hondt's later work

Ghent, attracted immediate attention and was later employed in various list systems of PR which were used extensively in Europe.

We can describe d'Hondt's scheme of election sufficiently by a single arithmetical example. Suppose that 3 parties A, B and C contest an election in which 5 seats are to be filled, and that A puts up 4 candidates, B 3 candidates and C a single candidate. Let the total votes polled by the candidates be:

party A ...... 5000 votes
party B ...... 4000 votes
party C ...... 1000 votes

	No. of votes per candidate		
No. of candidates	party A	party B	party C
1	5000	4000	1000
2	2500	2000	500
3	1666.67	1333.33	333.33

Figure 3.9

A scheme based on d'Hondt's proposal divides each party's vote by 1, 2, 3, ... to find the 'quota' needed to fill a single seat; and this quota, one of the figures on the right-hand side of the table of Fig. 3.9, is chosen so that the number of seats filled is just the total number available. In the example, choosing 1666 as the quota, A has three full quotas, B and two and C has none, so that A fills three seats, B two seats and A gets none, and the available seats are filled.

The table to find the d'Hondt quota is the same as that for constructing the diagram of Carroll's theorem; and *the d'Hondt quota, to within one vote, is the*

Exposé du système pratique de représentation proportionnelle adopté par le comité de l'Association Réformiste Belge (Ghent, 1885, p. 24) and his report at pp. 61--77 in *Conférence internationale pour la représentation proportionelle organisée par l'Association Réformiste Belge* (Brussels, 1885). A good description of list systems of PR is given by John H Humphreys, *Proportional Representation*, pp. 172--95. Cf. also Eward V Huntington, 'A New Method of Apportionment of Representatives', *Quarterly Publication of the American Statistical Association*, New Series, Vol. 17, September 1921, pp. 859--70, and 'Methods of Apportionment in Congress', *American Political Science Review*, Vol. 25, November 1931, pp. 961--5. 'Report upon the Apportionment of Representatives' by a committee consisting of C W Doten, E F Gray, W C Mitchell, E R A Seligman, A A Young and W S Rossiter, *Quarterly Publication of the American Statistical Association*, New Series, Vol. 17, December 1921, pp. 1004--13, F W Owens, 'On the Apportionment of Representatives' *Quarterly Publication of the American Statistical Association*, New Series, Vol. 17, December 1921, pp. 958--68.

height of the point H above the horizontal axis in the geometrical construction. The apportionment of seats on d'Hondt's scheme is that of Carroll's Theorem.

The theorem provides a rationale of the d'Hondt method. Suppose that, in an application of d'Hondt's method to PR, the elector is confined to voting for one or another of a given set of party lists of candidates. The allocation of seats is the same as that given by Carroll's Theorem. That is, the outcome is the same as would result if each voter were indifferent as between all candidates of the party he supports and preferred any candidate of this party to any candidate from another party, being indifferent again as between all candidates from these other parties; and if each party knew the strength of its own support and that of each of the others; and if each party directed the voting of its supporters with complete accuracy; and if each party decided 'rationally' on its voting strategy, in the sense of choosing in accordance with the maximin criterion. Put briefly, the outcome of the d'Hondt method is the same as that in which each elector has a certain restricted type of preference schedule, and each party has complete knowledge and acts in accordance with the maximin criterion. And the d'Hondt method secures this objective in a regime in which the knowledge of the parties is imperfect, in which the party's control over its voters is incomplete, and where the party has neither the knowledge nor the ability to act on the maximin criterion. The d'Hondt method imposes some degree of 'rationality' on a regime in which some of the requirements for rational action by the parties are not in fact satisfied.

Or again, in another application of the d'Hondt method, the elector is allowed to show his preferences as between the candidates belonging to a single party, the one which he supports, and the seats are allocated between the parties by taking the total number of votes for each party and finding the d'Hondt quota that gives entitlement to a single seat. This allocation is rational, or quasi-rational, in so far as it corresponds to what would happen if the parties knew their relative strengths, had complete control over the voting of their supporters and chose their voting strategies on the maximin criterion. The preferences that the electors feel for some candidates within the party to others, can be taken into account in deciding which of the party's candidates will get the seats available. Carroll's Theorem, of course, has nothing to do with the award of seats to particular candidates, but it provides a rationale for the total number of seats allotted to each party.

The justification of the d'Hondt method is, however, incomplete. The justification of the d'Hondt allocation of seats provided by Carroll's Theorem is incomplete in two respects. The theorem applies strictly only if the preference schedule of each elector is of a particular type, whereas in any election the schedules may diverge more widely or less widely from the type postulated by the theorem. (Likewise Huntington's criterion for a suitable allocation of seats, given in the ariticles that we allude to below, though we do not discuss this criterion, applies only to the extent that the schedule of each voter has the form of a plateau for the candidates of one party, the candidates of all the other parties being on a single lower plateau).

In another respect, the justification that the theorem provides for d'Hondt's system of PR may be incomplete: even granted all the assumptions of the theorem to be satisfied, the theorem may allocate seats to the parties in a way which is not itself ideal. This is indeed the case. It was pointed out by John H Humphreys, and was given a mathematical elaboration by Edward V Huntington, that the d'Hondt quota, or equally Carroll's Theorem, tends to give over-representation to the larger parties and under-representation to the smaller parties[29]. The effect of Carroll's Theorem, however, is not thereby undone. The allocation given by the theorem or the d'Hondt quota makes some approach to the ideal, without actually achieving it.

The games element, and with it the d'Hondt quota, come in via James Garth Marshall's pamphlet of 1853. Just before Carroll was writing, the d'Hondt quota (though without the name d'Hondt being attached to it) appeared in two pamphlets[30] which Carroll is likely to have read, and in a letter to *The Spectator*[31] which Carroll probably also read. It would seem, however that the d'Hondt quota and with it the games element, both of which are implicit in Carroll's Theorem, made their entry through James Garth Marshall's pamphlet *Minorities and Majorities: Their Relative Rights*. In various arithmetical examples relating to a two-party system, Marshall computes the minimum number of supporters that a party needs in order to fill 1, 2, 3, ... of the available seats. His work makes the implicit assumption that a party acts on the maximin criterion, and already in his tables of figures, the d'Hondt quota, as it was afterwards called makes its appearance. Carroll's four equations (see below, p. 159) put into algebraic form and generalize the treatment that Marshall had given.

Conclusion. Providing a rationale for the d'Hondt quota is of course no more than a side effect of Carroll's Theorem, though it does provide an instance of how, when expressed in mathematical form, apparently disparate elements may be found to be linked together. The theorem gives point to a suggestion which had originated as an empirical rule of thumb, by placing it as one element in a connected system.

---

29 John H Humphreys, *Proportional Representation*, pp. 178--80 and 188--90, Edward V Huntington, 'A new method ...' and 'Methods of apportionment ...'.

30 The pamphlets, *A Joint Candidate Scheme* by Frederick Seebohm and *The Joint Candidate Scheme for Constituencies Returning More Than Two Members* by James Parker Smith, were issued by the Proportional Representation Society, possibly at end-March or in April 1884: probably the only copies extant are those in the collection 'Letters 1884--1906' of the Electoral Reform Society.

31 Cf. letter by Charles T Mitchell in *The Spectator*, 15 March 1884.

3.8. The Fourfold Table and Carroll's Quota, *Q(s)*

Marshall's manner of proceeding was very like that of the mathematician of the seventeenth or eighteenth century who, without disclosing his own method, would issue a challenge to find the solution to a certain type of equation. Marshall had found the way to calculate the volume of support each of the two parties would need to fill a given number of seats in some prescribed constituency. His answers were invariably correct, but he did not disclose his manner of arriving at them. This seems to have been because of his weakness in algebra.

The algebra of Carroll's fourfold table may be regarded as Carroll's reply to the challenge of Marshall. It dealt with a broader problem in which Marshall's was a particular case. Carroll had first to see how Marshall would reach his answer. In turn our own effort has largely been to understand the fourfold table, having in this the help of the modern theory of the two-person zero-sum game.

Marshall uses his discovery to compare the two procedures which he terms the Single Vote and the Cumulative Vote. In, say, a five-member constituency, the Single Vote gives the elector one vote, while the Cumulative Vote gives him five votes which he can cast in any manner he chooses, giving possibly all five votes to one candidate, or else two votes to one candidate and three to another, and so on. And he shows that subject to certain assumptions, the two procedures will give the same answer; but he is at pains to show that there are considerations of a practical kind for thinking that the Cumulative Vote would give a higher degree of Proportional Representation than the Single Vote.

Carroll's theory, with results summarised in the fourfold table, covers all the possible cases, (a), (b), (c) and (d), that can arise in the family of procedures which he defines. This table enables him to work out the minimum percentage of votes that a party needs to fill any given number of seats in a constituency. These percentages are shown in his Table III. He then is able to construct his 'central argument'. one of no small sophistication, to prove that the procedure using $v = 1$ is superior to any of the others. He also shows that so far as pure theory goes, the Single Vote and the Cumulative Vote of Marshall, amount to the same thing.

> This Method [Cumulative Vote, referred to as the 'Method of Marks' by Carroll] would, I think be absolutely perfect, if only each elector wished to do all in his power to secure the election of *that candidate who should be the most generally acceptable*, even if that candidate should *not* be the one of his own choice: in this case he would be careful to make the marks exactly represent this estimate of the relative eligibility of *all* the candidates, even of those he *least* desired to see elected; and the desired result would be secured.
>
> But we are not sufficiently unselfish and public-spirited to give any hope of this result being obtained. Each elector would feel that it was *possible* for each other elector to assign the entire number of marks to his

> favorite candidate, giving all the other candidates zero: and he would conclude that, in order to give his *own* favorite candidate any chance of success, he must do the same for him.
>
> This Method is therefore liable, in practice, to coincide with [the Single Vote].[32]

Latterly Carroll usefully describes what can conveniently be called his 'practical scheme' of Proportional Representation. This employs the quota $Q(s)$ which features as case (a) in the fourfold table. But there is no question of using any other formula of the fourfold table, as a quota for the corresponding values of v and s. He has freed himself from the need to do this by showing that $v = 1$ gives a procedure which is superior to any of the others.

Apart from that, though he does not say so, he might well have considered any electoral procedure corresponding to (b), (c) or (d) of the fourfold table, as strengthening the power of the party machine and giving it more control over the party members. This he would have regarded as pernicious. As it is, his 'practical scheme' of Proportional Representation gives the utmost freedom to the individual voter and does not even require that a candidate should have party approval before standing as one of the official party candidates. This goes to an extreme in the setting aside of practical exigencies. Marshall was more realistic and stressed the need that party candidates should stand only with party approval. The two writers took different views of the optimum procedure for the parliamentary election. Carroll got the help of extreme simplicity by proposing that the number of seats awarded to a party should be in accordance with $Q(s)$, his variant form of Droop quota: a party's total vote was to be got by adding up the votes of its individual candidates and the number of seats awarded would then be in accord with $Q(s)$. He does so in such a way as to give the party machine not a shred of additional power. Marshall envisaged the m seats available as being allotted to the m candidates with the highest numbers of votes in the election. This proposal is looser than the answer proposed by Carroll.

Carroll's disregard of ties in the voting. Neither in his discussion of the fourfold table nor at any other part of the booklet, does Carroll refer to the possibility of ties in the voting. Sometimes a tie in the voting will be of a purely formal nature and will not affect the decision reached. For example, for $m = 5$ the optimum strategies of the two parties may be A {51, 50, 0, 0, 0}, B {50, 50, 50, 0, 0}. A tie would occur in the voting, but the election result {2;3} is never in doubt. But a tie may occur which leaves the election result in doubt. Suppose that in a 3-member constituency, the optimum strategies for the parties are A {71, 70, 0}, B {70, 70, 0}. For the parties

---

32 From *A Discussion of the Various Methods of Procedure in Conducting Elections*, 1873, reprinted in Duncan Black, *The Theory of Committees and Elections* (Cambridge: Cambridge University Press 1958), p. 218, emphasis in original.

using these strategies the outcome can be stated as {1 seat plus 1 tie in a 3-way tie; 0 seat plus 2 ties in a 3-way tie}.

For a 'one-shot event' like this, however, the laws of mathematical probability and of mathematical expectation do not apply, and we cannot deduce what a party's reaction would be: it might choose to use the strategy which would give rise to the tie or it might not.

Proof that both parties will act on the maximin principle. In order to find out the best strategy for it to play a party need consider only its optimum strategies; and for each party the party's choice may be confined to the choice of one strategy out of m optimum strategies which are open to it. A party may choose one strategy out of a much larger number than this (indeed it may have open to it, the choice of one out of many thousands of strategies): but by one out of the m optimum strategies we have defined the party can fill the maximum number of seats which it is open to it to fill. It has for instance, no more efficient way of filling, say, two seats ($m \geq 2$) than by putting up exactly two candidates and giving each, so far as possible, the same number of votes.

Not only does the party A, say, knew an optimum strategy for filling one or two, ...or m seats, but it knows the optimum strategy for B filling one, or two, ..., or m seats. Granted, of course, as we assume to be the case, that each party knows the number of its own supporters and the number of supporters of its rival.

With this knowledge, a party knows an optimum means by which it can fill say three seats in a five-seat constituency. It also knows an optimum means which its rival has open to it of filling three seats or four or five seats. A can calculate with little difficulty whether it would, using this optimum strategy, be able to fill three seats, or whether its rival would be able to prevent it from doing so. And if it finds that it would be able to fill three seats it can proceed further and calculate against the best-directed efforts of its rival, it would be able to fill four seats, and so on.

The outcome of the process of calculation of this kind, is that each party is able to calculate the maximum number of seats which it would be able to fill against the most efficient counter-strategy which its rival has open to it to fill. It will, therefore, acting rationally, choose to maximise the number of seats it fills against the best-directed strategies which it is open to its rival to choose. Put in another way, the party, acting rationally, will choose to maximise the minimum number of seats to which the action of its rival restricts it. But this is simply to say that in a contest of this kind, each party will act on the maximin criterion.

Knowledge of the kind we describe, is in no way confined to 'the party' concerned or its officers or members. Granted that the number of supporters of each party is known, and that the procedures in use will be that defined as we presently suppose, by $v = 1$, knowledge of this kind about the maximum number of seats a party can fill and of *one* of the policies by which it can so (there may be other policies too, open to the party to achieve the same result) is *public* knowledge and equally accessible to all who understand how this electoral procedure works.

Moreover the same is true for each of the other procedures in the family of procedures which Carroll considers, defined by the different values of v, $1 \leq v \leq m$.

Bearing this in mind we may simply refer to a party choosing its strategy on the maximin principle, or in accordance with the maximin procedure. Each party will then use either the particular strategy which we have defined (in the example we have chosen, putting two candidates, each with the same numbers of votes, to within a single vote).

If we assume complete knowledge of these matters, on the part of all electors, there can be no disagreement among different electors as to which value of s is appropriate to a party to use. (We assume, like Carroll, that no ties can occur in the voting.)

Furthermore 'wastage in voting' can occur only through lack of knowledge on the part of voters of how other electors are casting their votes. If we assume complete knowledge on the part on the voters as to the numbers of electors who are casting their vote for the various candidates, no candidate will be given any more votes than would be needed to secure his election.

Number of candidates. Now granted this knowledge of the optimum strategy for each party, and the optimum number of seats for a party to aim to fill, there goes also the knowledge of an optimum number of candidates for each party to put up. A party's supporters, we may assume, will wish in most circumstances to concentrate their votes on the s candidates who would fill the s seats that the party aims to fill - and indeed as the suppositions we make, on the s seats which everyone who understands the process *knows* the party will be able to fill.

Sometimes by a special device, it will make no difference to the outcome of the election whether the supporters of the party 'waste' their votes or not; indeed the device may manage to make it impossible for them to waste their votes. This is true of the 'practical scheme' of election, as we term it, devised by Carroll. In it all the votes cast for a party's candidates are collected up and pooled, and the number of seats awarded to a party is then equated to the number of complete Droop quotas (in the Carroll variant of the Droop quota) cast in favour of the party's candidates.

***Q(s)* a function of the number of seats to be filled.** Carroll modifies the Droop quota so as to obtain a form specially adopted to the two-party system. His argument assumes votes to be infinitely divisible, and makes the set of assumptions which today we associate with use of the maximin criterion. Each of the parties A and B is assumed to know the number of its own supporters and that of its rival, and to be able to direct the votes of its supporters on to such candidates as it chooses.

In a constituency with e voters and m members, a *sufficient* condition for A to fill s seats, is that it should be able to give more votes to each of s candidates, than B can give to each of $(m + 1 - s)$ candidates. Suppose A has x votes, so that B has $(e - x)$. A party with x votes, assumed to be infinitely divisible, which aims to fill s seats, will have (by J G Marshall's Theorem) as an optimum strategy, giving exactly x/s votes to each of s candidates. Likewise if B, with $(e - x)$ votes aims to fill $(m + 1 -$

s) seats, an optimum strategy for it will be to give exactly $(e-x)/(m+1-x)$ votes to each of $(m + 1 - s)$ candidates. Carroll supposes each party to act on the assumption that whatever strategy it chooses, its rival will employ the most efficient counter-strategy. If so, a *sufficient* condition for A to be able to fill *s* seats is that

$$\frac{x}{s} > \frac{e-x}{m+1-s}$$

$$x > \frac{se}{m+1}, \quad 1 \leq s \leq m$$

The number of votes *necessary* and *sufficient* to enable A to fill *s* seats will be got by taking *x just* greater than this amount. Since *x* must be an integral number of votes, the *necessary* and *sufficient* condition for a party with *x* votes to fill *s* seats is that

$$x = \frac{se}{m+1} + i, \quad 0 < i \leq 1$$

where *i* is chosen to give the integer next greater than the quotient.

The number of votes concerned is a function of *x*, but will not as a rule be a linear function. Carroll himself denotes this number of votes by *Q*, but it seems better to elaborate a little and we will denote the number of votes necessary and sufficient to enable a party, acting on the maximin criterion, to fill *s* seats, by

$$Q(s) = \frac{se}{m+1} + i, \quad 0 < i \leq m+1$$

or, more usually, adopting a notation similar to that for the Droop quota,

$$Q(s) = \frac{se}{m+1} + \frac{t_s^*}{m+1}, \quad 1 \leq t_s^* \leq m+1$$

To illustrate, $e = 686$, $m = 5$ gives $Q_d = 115$, $t = 4$, $e > mt$.

$Q(1) = 115$, $t_1^* = 4$; $Q(2) = 229$, $t_2^* = 2$; $Q(3) = 344$, $t_3^* = 6$; $Q(4) = 458$, $t_4^* = 4$; $Q(5) = 572$, $t_5^* = 2$, where t_1^* is the value of *t* associated with *Q(1)*, t_2^* the value of *t** associated with *Q(2)*, and so on.

The interpretation of these figures (assuming Carroll's analysis is correct) is that given two parties, each acting on the maximin criterion, 115 is the smallest number of votes that enables a party to fill one seat, and 229, just under twice the amount enabling it to fill one seat, is the smallest number of votes that enables a party to fill two seats. With any number of votes between 115 and 228 inclusive, the party would fill one seat. The other values of *Q*(*s*) are to be interpreted in the same way.

Using *Q(s)*, a procedure cannot fill more than *m* seats. To fill more than *m* seats would require the procedure to award *s* seats to the one party and say, $(m + 1 - s)$ seats to the other. The one party would have to hold *Q(s)* votes or more, and the other $Q(m + 1 - s)$ votes or more. But this would require as a minimum, a total number of votes

$$Q(s)+Q(m+1-s)=\frac{se}{m+1}+i+(m+1-s)-\frac{e}{m+1}+j, \quad 0<i\le 1, \quad 0<j\le 1$$

The quota $Q(s)$, therefore cannot fill more than the m seats available, and satisfies the condition we accepted as fundamental in defining the Droop quota - as indeed is evident from the manner in which Carroll arrives at the formula.

With $Q(s)$ in use, an increase in the size of A, B remaining constant, cannot reduce the number of seats filled by A, and no effect of the type of the Droop paradox is possible. Proof. Suppose that initially, for $A = A_1$, $B = B_1$, $e_1 = A_1 + B_1$, A qualifies to return h candidates, $0 \le h \le m$. Obviously the proposition will hold for $h = 0$.

The quota needed to return h candidates is

$$Q(h)=\frac{he_1+t_h^*}{m+1}, \quad 1\le t_h^* \le m+1$$

It is required to prove that when the number of A-votes increases by say k, $k \ge 1$, to $(A_1 + k)$, and e increases to $(e_1 + k)$, A will be able to return at least h candidates; that is, it is required to prove

$$A_1+k \ge Q(h) \quad \text{for } e=e_1+k$$

$$\ge \frac{h(e_1+k)+t^*}{m+1}, \quad 1\le t^* \le m+1$$

$$(m+1)(A_1+k) \ge he_1+hk+t^*$$

Since $t^* \le (m + 1)$, a sufficient condition for this will be

$$(m+1)(A_1+k) \ge he_1+hk+m+1$$

$$\ge he_1+t_h^*+(m+1)k$$

$$\ge he_1+1+mk+k$$

$$\ge h(e_1+k)+m+1, \quad 0\le h\le m+1$$

since $k \ge 1$. Hence the proposition.

Some algebraic relations governing Q(s) and the Droop quota.

(*a*) *$Q(1) = Q_d$ and the two are logically equivalent.*

This follows since the algebraic expressions for the two are identical.

The concept of the Droop quota applies if and only if $e \ge mt$, and (*a*) shows that $Q(1)$ and Q_d are logically equivalent. This *suggests* that for Carroll's concept to apply, again the requirement $e \ge mt$ must be satisfied and *we will assume without proof that this is so.*

(*b*) *For $t_1^* = 1$, $Q(s) = sQ(1)$, $1 \le s \le m$.*

For $t_1^* = 1$,

$$Q(1) = \frac{e}{m+1} + \frac{1}{m+1}$$

$$sQ(1) = \frac{se}{m+1} + \frac{s}{m+1}, \quad 1 \le s \le m$$

Since $Q(1)$ is an integer, so also $sQ(1)$ must be an integer; and since $s/(m+1) < 1$, $sQ(1)$ must be the integer next above $se/(m+1)$, that is, must be $Q(s)$. Hence for $t_1^* = 1$ $Q(s) = sQ(1)$, and in particular $Q(m) = mQ(1)$.

(c) Except for $t_1^* = 1$, $mQ(1) > Q(m)$.

$$Q(1) = \frac{e}{m+1} + \frac{t_1^*}{m+1}$$

and let us suppose $2 \le t_1^* \le m + 1$. Then

$$Q(1) \ge \frac{e}{m+1} + \frac{2}{m+1}$$

$$mQ(1) \ge \frac{me}{m+1} + \frac{2m}{m+1}$$

while $$Q(m) \le \frac{me}{m+1} + \frac{m+1}{m+1}$$

$$< mQ(1) \qquad \text{for } m+1 < 2m \Rightarrow m > 1$$

(d) For both $t_1^* = m$ *and* $t_1^* = (m + 1)$, $Q(s) = sQ(1) - (s - 1)$, *and in particular* $Q(m) = mQ(1) - (m - 1)$.

First, for $t_1^* = m$,

$$Q(1) = \frac{e}{m+1} + \frac{m}{m+1} \qquad \text{and is integer}$$

$$2Q(1) = \frac{2e}{m+1} + \frac{2m}{m+1} \qquad \text{and is integer}$$

$$= \frac{2e}{m+1} + 1 + \frac{m-1}{m+1} \qquad \text{and is integer}$$

$$= \text{integer next above } \frac{2e}{m+1} + 1$$

But $Q(2)$ = integer next above $2e/(m+1)$.

$$2Q(1) = Q(2) + 1$$
$$Q(2) = 2Q(1) - 1$$
$$3Q(1) = \frac{3e}{m+1} + \frac{3m}{m+1}$$
$$= \frac{3e}{m+1} + 2 + \frac{m-2}{m+1}$$
$$= \text{integer next above } \frac{3e}{m+1} + 2$$

But $Q(3)$ = integer next above $3e/(m+1)$

$$\therefore \quad 3.Q(1) = Q(3) + 2$$
$$Q(3) = 3Q(1) - 2.$$

We see that in general $Q(s) = sQ(1) - (s - 1)$, and in particular $Q(m) = mQ(1) - (m - 1)$.

Second, for $t = m + 1$,

$$Q(1) = \frac{e}{m+1} + 1 \text{ and is integer.}$$
$$Q(2) = \text{integer next above } \frac{2e}{m+1}.$$

Since $(2e/(m+1)+2)$ is integer, the integer next above $2e/(m+1)$ is $2e/(m+1)+1$ = Q(2) = 2Q(1) - 1.

Similarly the integer next above $3e/(m+1)$ is $3e/(m+1)+1$ = Q(3) =3Q(1) - 2

In general, for $t_1^* = (m + 1)$, $Q(s) = sQ(1) - (s - 1)$. In particular, $Q(m) = mQ(1) - (m - 1)$.

This pair of relations indicate the maximum economy of votes from using $Q(s)$ instead of the Droop quota in the election of candidates. Zero economy is got for $t_1^* = 1$.

Under (a) it was observed that the algebraic expression for $Q(1)$ is the same as that for the Droop quota. Taking into account that the algebraic expression for Q_d is valid for the range $e \geq mt$, we decided to assume that the formula for $Q(1)$ will also be valid for this range of the variables.

From (a) - (d), $Q(s) \leq s\,Q_d$, $1 \leq s \leq m$. This linkage between the Carroll quota and the Droop quota suggests, without proving it, that restricting the variables to the range $e \geq mt$ is sufficient to ensure the validity of $Q(s)$.

Notes on the following examples. If the parties hold between them m complete Droop quotas, the Droop quota will fill the seats available. Since $Q(s) \leq s\,Q_d$, $Q(s)$

too in these circumstances will fill all the seats and assign the same number of seats to each of the two parties. No example is needed to illustrate this feature.

$Q(s) \leq s\,Q_d$ and sometimes when only $(m - 1)$ complete Droop quotas exist, $Q(s)$ will fill all the seats available instead of only $(m - 1)$. This is illustrated in example 1.

It may be, however, that where only $(m - 1)$ complete Droop quotas exist, $Q(s)$ will fill only $(m - 1)$ of the available seats, though instances of this kind appear to be rare. In example 2, $e > mt$, and in example 3, $e = mt$, and in neither case does $Q(s)$ fill all the available seats.

A further possibility to be borne in mind is that a party may use neither the Droop quota nor $Q(s)$, but will take into account the optimum strategies open to itself and to its rival, and will use the payoff matrix to select its maximin strategy. This raises the question whether the allocation of seats got in this way may differ from that given by $Q(s)$.

Carroll's algebraic derivation of $Q(s)$ presupposes continuous quantities, and from the side of theory the answer to the query would seem to be open. With this in mind we have experimented with a number of examples, and we have found no instance in which a *definite* allocation of seats got from the payoff matrix (without recourse to a tie-breaking mechanism) differs from that given by $Q(s)$.

Of course if the matrix is used but supplemented by a tie-breaking mechanism, the payoff matrix itself cannot be said to yield any definite allocation of seats. No sensible comparison is possible between the 'outcome' of the payoff matrix and that given by $Q(s)$.

Example 1. Let $e = 100$, $m = 3$, $Q_d = 26$, $t = 4$, and $e > mt$. $Q(1) = 26$, $Q(2) = 51$, $Q(3) = 76$.

For A = 76, B = 24 only two complete Droop quotas exist. A procedure using the Droop quota would fill only two seats, but $Q(3)$ awards the three seats available to A.

Constructing the payoff matrix would show that with the maximin strategy A{26, 25, 25} A is bound to fill the three seats.

Example 2. Let $e = 100$, $m = 3$, $e > mt$ and so on, as in Example 1, but take A = 75, B = 25. $Q(s)$ will fill only two of the three seats, awarding them to A.

The payoff matrix would show the maximin strategies as A{25, 25, 25} and B{25, 0, 0}, giving rise to a tie. A would fill either two or three seats, and B fill either one seat or no seat.

Example 3. $e = 12$, $m = 3$ gives $Q_d = 4$, $t = 4$, $e = mt$. $Q(1) = 4$, $Q(2) = 7$, $Q(3) = 10$.

For A = 9, B = 3 only two complete Droop quotas exist and using the Droop quota A and B between them would fill two seats. Using $Q(s)$ A fills two seats and B fills no seat, leaving one seat unoccupied.

The payoff matrix would show B is bound to use the maximin strategy B{3, 0, 0}, to which the appropriate counter-strategy is A{3, 3, 3}. A would fill two seats and tie with B for one seat.

Example 4. $e = 91$, $m = 5$ gives $Q_d = 16$, $t = 5$, $e > mt$. $Q(1) = 16$, $Q(2) = 31$, $Q(3) = 46$, $Q(4) = 61$, $Q(5) = 76$.

For A = 46, B = 45, A gets two complete Droop quotas, and B two complete Droop quotas and only four complete Droop quotas exist. $Q(s)$ awards three seats to A and two to B, and fills all the available seats.

In the payoff matrix for the optimum strategies for the parties, one cell shows a five-way tie, another a four-way tie, and another a three-way tie. The matrix fails to show any definite choice of strategies by the parties.

Q(s) more important in the 'small' constituency with a 'large' number of seats. In any constituency ($e \geq mt$) the probability of only (m - 1) complete Droop quotas being shared between the two parties is $m(t-1)/(e-1)$. In the parliamentary constituency the number of voters will run to thousands. The probability of the occurrence of only (m - 1) complete Droop quotas may well be judged to be negligible. But in the 'small' constituency in which the number of seats is comparatively 'large', the probability of the occurrence of only (m - 1) complete Droop quotas may be appreciable. $Q(s)$ may give a result while the Droop quota does not do so, and in the small constituency the advantage of using $Q(s)$ will be greater than in the large.

In the event of both the Droop quota and $Q(s)$ failing to give an answer, as in examples 2 and 3 above, recourse may be had to the maximin solution, got from the use of the payoff matrix supplemented by a tie-breaking mechanism.

Conclusion. A serious drawback is that under Carroll's $Q(s)$, the everyday conception of the quota as a price per seat, no longer applies. We lose much of the benefit of common-sense discourse: a mode of speech adapted to the quota in the everyday sense becomes incorrect, and this opens up a broad path to confusions of thought. Again the economy which $Q(s)$ seems to offer, of enabling a party to fill a given number of seats, with a smaller number of votes, is to some extent illusory: under the operations of Game Theory the party might be unable to fill as many seats as $Q(s)$ promises: or on the other hand, the party might fill more seats than $Q(s)$ entitles it to.

The advantages of $Q(s)$ would seem unable to compensate for disadvantages which amount to errors.

3.9. Carroll's Practical Scheme and the Single Transferable Vote

Regardless of the quota used, a candidate who wins a seat will, as a rule, have votes which are surplus to his requirement for election; and again a candidate may have votes which don't entitle him to election. Recall that Carroll's suggestion is that the candidates of a party shall exchange these surplus votes among themselves, so as to take up the number of seats *to which the party is entitled.*

This system of PR allocates seats to the two parties in some rough proportion to their strengths. Also by giving precedence to the candidate with more votes over the candidate with fewer, it gives the electors some degree of control over which candidates will take up the seats to which a party is entitled. The scheme commends itself to common sense and before going on to consider parts of it in more detail, we first discuss the criticisms which Carroll makes of the scheme which, at the time, had widespread support in the country, the single transferable vote (s.t.v.).

Criticism (i): the difficulty of forming and writing down a preference ordering. Earlier, Walter Baily, by whose work Carroll was greatly influenced, had commented:

> In cases where [the s.t.v.] could be conveniently used, such as electing the committee of literary and scientific societies, clubs, institutes, libraries &c., no better system could be devised for obtaining a thoroughly representative body; but I am disposed to think ... that there would be a good deal of inconvenience in using it for large constituencies where the voters would be counted by thousands, and would be but too often imperfectly educated... Many a voter would get hopelessly confused.[33]

In his article 'Parliamentary Elections', *The St. James Gazette*, 5 July 1884, [Carroll] puts this objection quite bluntly:

> That the process of marking a ballot-paper should be reduced to the utmost possible simplicity, to meet the case of voters of the very narrowest mental calibre, I should have put as an axiom, but that the [Proportional Representation] Society appears to ignore it. No doubt they have found many school-children able to tick off, with great readiness, lists of kings and conquerors in a supposed descending scale of merit - but try it on Hodge straight from the plough! Give him a list of half-a-dozen of the neighbouring farmers, to be arranged in their order of merit, and see if he will ever be able to make up his mind! 'I know who's best of *two*', he might

33 Walter Baily, *Proportional Representation in Large Constituencies*, p. 9.

tell you, 'but blessed if I can say who's first, and second, and third, and fourth!'

...Let each Elector vote for one candidate only.

In the corresponding passage in the booklet the tone is muted:

> 'The Proportional Representation Society' proposes to let each Elector hand in a list of Candidates, marked in the order of his preference; and that his vote, if not required for his No. 1, should be transferred to his No. 2, and, if not required for him, then to No. 3, and so on. One great objection to this method is the confusion it would cause in the mind of the ignorant Elector, who, though quite able to name his favourite Candidate, would be utterly puzzled if told to arrange 5 or 6 names in order of merit.

To estimate the force of this objection one must take into account the great extension of the franchise on which the country had embarked, giving the vote for the first time to a wide class of citizens. Besides this the increase in the size of the parliamentary constituency would magnify the number of candidates and increase the difficulties of the elector in forming an order of preference among them. Obviously in the circumstances of the day, Carroll's objection would have a special force.

Quite apart from such exceptional circumstances, anyone confronted with half a dozen names, whatever his education or experience, is going to find it difficult to arrange them in order of preference. Ask him a few days later, to order the same names and he may give quite a different answer. Any ordering he makes is likely to be to some degree fictitious. Indeed Carroll had earlier taken notice of this feature in his theory of the committee. In a procedure designed to deal with important topics, the members were to be asked to submit complete strong orderings of the proposals under consideration. If these lists showed no motion able to get a simple majority against each of the others, the chairman was to enquire whether certain members would be willing to have their lists amended, so that with the minimum amount of change in this way, a majority motion might be enabled to emerge.

This procedure had been intended for use by a Christ Church committee: but Carroll is against the use of preferential lists for the elections of members to parliament: instead, the voter will name only the candidate of his first preference. And the question arises: Would a party be entitled, on the basis of the very primitive information, to collect up all the votes which had been cast for its candidates?

The answer arises out of the view Carroll takes about the preference schedules of the elector. Each elector who goes to the poll, he assumes, prefers *every* candidate of the party he favours, to *any* candidate put up by the other party, as in the set of schedules reproduced in Fig. 3.10 below. The A-supporter, say, prefers every A-candidate to any B-candidate, *irrespective of what his estimate of the merits of the A-candidates or B-candidates as individuals may be.* He assumes the elector to cast his vote essentially for a party, and to be relatively indifferent as to which of the party's candidates take up the seats awarded to it. On this view the procedure allowing a party to collect up the votes given to its candidates would seem to be justified, for these votes had been intended essentially for the party rather than the individual candidate. At the same time, the second of his procedures (by which the candidates awarded the seats will be those with the most votes) is an effective way of taking into account the preferences of the voters in relation to the candidates themselves. It caters for the voters' preferences with the minimum of fuss and effort.

Carroll's procedure and the s.t.v. each give the party the same number of seats. To make a straight comparison between Carroll's scheme and the s.t.v., let us assume that each voter has in fact a complete schedule of preferences in regard to all the candidates who stand in his constituency. We retain the assumption of a two-party system and can safely assume that each party will put enough candidates to fill all the seats it will be entitled to. Under Carroll's scheme the voter names only the candidate of his first preference: under the s.t.v. he submits a complete list of preferences. Both schemes use the Droop quota.

On Carroll's scheme, suppose that the party A puts up four candidates. If A gets 2.8 Droop quotas of votes, say, it is awarded two seats, and these are taken up by the two A-candidates getting the highest numbers of votes.

Under the single transferable vote, each schedule of an A-supporter will have the four A-candidates in the top places, above any of the B-candidates. In all there will be 2.8 Droop quotas of these schedules. Before the transfer of a single vote can be made from a schedule of this kind to a B-candidate, all the 2.8 Droop quotas of votes from these schedules must first be distributed among the A-candidates; and none of this 2.8 Droop quotas of votes can be transferred to a B-candidate, until it has been found that it cannot be utilized by an A-candidate to gain election. Hence under the s.t.v. the A-party will fill two seats, the same number as under Carroll's scheme. Thus the number of seats filled by the two parties will usually be the same under both schemes. Since Carroll's quota $Q(s)$ may be smaller than sQ_d, it may fill m seats in cases where the Droop quota fails to do so.

Carroll's criticism of the s.t.v. In a two-party system Carroll's scheme and the s.t.v. will almost always give the parties the same number of seats, but it is still possible that the two procedures will diverge on the particular candidates they will choose to fill the seats. Soon after the second edition of *PPR* was completed (preface dated 1 Jan. 1885) Carroll began thinking again about the s.t.v. Already in May 1884 he had attempted to find examples in which the s.t.v. would give the wrong answers, and now he resumed this line of thought. He needs to show at this stage not only that the s.t.v. may choose different candidates from his own scheme (which may be fairly easy), but also that it may do so in circumstances in which the candidate chosen by his own scheme would clearly be those that fit the situation. The diary entries are[34]:

---

34 Cf. Roger Lancelyn Green (ed.) *The Diaries of Lewis Carroll*, vol. ii.

Jan. 16 (F). 1885. After two days of experiments, I have at last constructed a clear crucial case of failure for the Proportional Representation Society method, which I think of circulating as a supplement to my pamphlet.

Jan. 22/23. 1885. I have been so long with Logic that I have not finished above paper.

Feb. 2 (M). 1885. Ordered of Baxter 1,000 copies of supplement to *Parliamentary Representation*, to be posted to all MPs, and members of the 'Prop. Rep. Soc.'

Feb. 11 (W). 1885. Sent to Baxter the MS of a Postscript to the Supplement to meet an objection of Mr. Courtney's.

March 29 (Sun). 1885. [He lists among his projects on hand] new edition of *Parliamentary Representation*, embodying supplement, etc.

An example, and 'the Edgeworth principle'. Apart from polemics, the *Supplement* to the booklet and the *Postscript to Supplement* are taken up by discussion of the example of Fig. 3.10. In it we have replaced Carroll's names of contemporary politicians by symbols, and on the last schedule the symbols a_1, a_2, a_3, a_4 may be arranged below b_1 in any order.

a_1	a_2	a_1	a_4	a_3	b_1
a_2	a_1	a_4	a_1	a_1	
a_3	a_3	a_2	a_2	a_2	
a_4	a_4	a_3	a_3	a_4	
b_1	b_1	b_1	b_1	b_1	
3030	2980	2020	1100	790	2070

Figure 3.10

For a three-member constituency, with 11999 votes in all, the Droop quota needed to return a single member is 3000.

Under Carroll's procedure the votes are:

$$a_1, 5050; a_2, 2980; b_1, 2079; a_4, 1100; a_3, 790.$$

a_1, a_2, and a_4 'would then club their votes, making 9130 votes, which would suffice to return all three' (*Supplement*, p. 172).

For this group of schedules the s.t.v. chooses a_1, a_2, and a_3. It would exclude the candidate a_4, who 'most undoubtedly ought to be returned' (*Supplement*, p. 171), though Carroll at this does not explain why this should be so.

The objections of Leonard Courtney and others to this opinion, led him to formulate this justification for holding the triplet a_1, a_2, a_4 to provide a greater degree of PR than a_1, a_2, a_3. (In the next quotation we again replace names of politicians by symbols).

> The 6010 electors, who put a_1 and a_2 as their first two favorites, have a clear *moral* right to return these two, as they number more than two full quotas, and it is a mere accident (which they would have avoided had they known how the voting was going on) that they did not divide themselves so as to *secure* this result.
>
> Now suppose them to have exercised this right, and that a_1 and a_2 are returned. Then the 3120, who put a_1 and a_4 as their first two favorites, would undoubtedly (a_1 now being safe) all vote for a_4. *He* therefore has also a moral right to be returned. (*Postscript to Supplement*, pp. 173--4).

The making of a single wrong choice is of little account and what matters is the *proportion* of wrong choices made by a procedure. But to justify his view that, by comparison with the s.t.v. his own scheme chooses the right candidate, Carroll was led to formulate the principle of coalition-formation.

We are asked to envisage the electors as being free to give effect to their preferences by forming coalitions, with agreement by one Droop quota of the electors as the price of filling one seat, agreement by two Droop quotas as the price of filling two seats, and so on. The same principle of coalition-formation is precisely the notion which Edgeworth takes as fundamental[35]; it is a basis for the Theory of Games; and also for some parts of Carroll's booklet.

Summing up the comparison. To enable the Droop quota to operate, the voters must be arranged into the appropriate 'parties' or groups. The s.t.v. proposes, on the evidence presented in their schedules, to arrange a succession of groupings. The Droop quota can then be put to work on the groupings which have been discovered, or created, or deemed to exist. By contrast Carroll's scheme is intended to operate in an environment in which each voter is known to fall into one or other of two political parties. No effort need be made to find the proper groupings: they are self-evident and the Droop quota can be put to work directly.

We have not attempted to discuss the extent to which Carroll's scheme would in fact achieve PR, beyond showing that in a two-party system it would do so just as well as the s.t.v. Of course it would work in a way which was far better understood in the community concerned.

---

35 F Y Edgeworth, *Mathematical Psychics* (London, Kegan Paul, 1881, facsimile reprint LSE Reprints of Scarce Tracts in Economic and Political Science no. 10, 1932). [Black probably intended to refer to Edgeworth's remarkably prescient discussion of combinations and co-operation on pp. 43--50].

In circumstances like those of Northern Ireland, its use might well be contemplated.

3.10. Allocating Members to Districts

The Droop quota leads to a change of plan. No section of the booklet would give more difficulty to the reader than its discussion of the allocation of Members to Districts. (Our discussion will retain the initial capitals Carroll gives to some of his nouns. Possibly he does so as a reminder that a 'District' is a potential district, not yet defined geographically or otherwise; an 'Elector' is a potential voter, legally qualified to cast a vote though he may not turn out at an election to do so; and a Member is a potential member of the 'House'.) The article 'Parliamentary Elections', *St James's Gazette* of 5 July 1884, proposes that a District should be given a number of Members in proportion to its number of Electors:

> It seems clear that each district (*sic*) should return several Members, so that Minorities may have a chance of returning some. But, if this be so, there is no reason why the districts should be *equal*, provided only that the number of Members returned be proportioned to the number of Electors in the district.

In the article 'Redistribution' of 11 October 1884, and again in the booklet, he is forced to abandon this proposal. He has now shown the desirability of giving the Elector a single vote in a multi-Member District, and is aware that this will be equivalent to a procedure which makes use of the Droop quota. Consider, then, the outcome of giving the districts number (1), number (2) and number (3), with say 10,000, 20,000 and 30,000 Electors, one, two and three Members receptively.

For number (1), 1 Droop quota = 1 × 5001 = 5001

For number (2), 2 Droop quotas = 2 × 6667 = 13,331

For number (3), 3 Droop quotas = 3 × 7501 = 22503

But the Droop quota provides a measure of the number of Electors represented, and the percentages of Electors represented in the three Districts are 50+%, 67% and 75% respectively. Allocation of seats in accordance with number of Electors is seen to yield a nonsensical arrangement.

Abandoning his earlier scheme, Carroll now decides that the equitable arrangement is to award one Member for each Droop quota of Electors in the District concerned. If, as is quite easy, we then arrange that the Droop quota shall be of the same size throughout the kingdom, then in each District the individual Member will represent the same number of Electors. Obviously this will achieve a highly desirable objective.

But this manner of allocation too, gives rise to awkwardness. Consider three Districts which have one, two and three Members respectively, and take into account that each District will have one Droop quota of 'unused votes' or of 'unrepresented Electors'. The first District will have in all two Droop quotas of Electors, of whom one Droop quota will be unrepresented. The second District has in all three Droop quotas of Electors, with one Droop quota unrepresented; and the third District has four Droop quotas of Electors, of whom one Droop quota is unrepresented. The percentages unrepresented in the districts are 50%, 33% and 25% respectively, and the smaller the District, the more harshly it is treated.

In spite of this obvious weakness, Carroll thought it desirable to allocate Members after the manner dictated by the Droop quota. This was inevitable if each Member in the House were to represent the same number of Electors. His earlier predilection had been to proportion the number of Members to the number of Electors in the District and he was aware that this would continue to be the predilection of most readers. In the second edition of the booklet he transferred most of the argument to an appendix.

Members must be allocated to parliamentary Districts on the basis of the number of potential voters not actual voters, for the allocation is made at a time before any election is held as, as Carroll says; 'Before the poll is closed' there are no actual districts but only potential districts. Likewise the Candidates for whom we may envisage the Electors as voting are potential and not actual candidates. In his theory of the allocation of Members to Districts Carroll reaches the stage at which he has to close the circle of reasoning and have his system as a self-consistent whole, deriving from a limited set of premises. By this stage Carroll has used two distinct approaches to the problem of proportional representation and each has led to the same conclusion. Following Baily he has sought to achieve proportional representation by minimising the number of unused votes or as this becomes in his formulation 'the number of voters unrepresented'. This leads to the use of the Droop quota in an election. Again the J G Marshall and Games line of thought leads to the same electoral procedure: the voter is to be given a single vote and a decision arrived at by a procedure which is equivalent to the use of the Droop quota.

The theory of allocation of members, therefore, can take it that the procedure in use will be the Droop quota. Also, before the poll is closed it seems sensible to suppose that all the Electors of each District will go to the polling booths to cast their votes, despite the fact that when the election is actually held, some of them won't vote: the simplest thing is to assume that everyone with the legal right to vote will so.

This is the background and consideration of the allocation of members allows Carroll to close a gap which has existed in his theory. Even though the Droop quota is to be used, it might be that while in one District, a quota of say 1,000 might be needed to fill one seat, in another District the quota needed would be say 100,000, and, as pointed out above, the number of electors in the English constituencies, showed enormous variation. Now to close his system of reasoning, and ensure proportionality of representation as between one District and another, Carroll requires that each Member shall represent their own number of Electors. The number of Electors represented by the Member for any District will be the Droop quota for that District. The Droop quota, therefore, must be the same for all Districts.

[Black's manuscript breaks off part way through the next sentence. We continue with the exposition he clearly intended, and add our own evaluation of this part of Carroll's argument.]

By a piece of algebra, he shows how this minimum size of quota is to be obtained [(*PPR*, chapter II, §§ 3 and 4). Where M = the size of the House, E = the number of electors, and D = the number of districts, the (Carroll) quota $Q = E/(M + D)$. Where m = the number of seats in a given district and e = the number of electors in that district, Carroll's algebra gives an apportionment of seats to each district of $m = (e/Q) - 1$.

Carroll next puts E at 5,000,000, M at 660, and D at 180 (each district thus returning on average between three and four MPs). These were realistic numbers. (In the General Election of 1885, the electorate E was 5.7 million including Ireland, or 5.0 million excluding Ireland. The size of the House M was 670 including Ireland, and 569 excluding Ireland. The actual number of districts D was 542, but this was a consequence of the division of the country into one- and two-member districts in the Redistribution Act, and the consequential abandonment of the limited vote.[36]) These numbers enable him to calculate both the Carroll quota and what might be called the natural quota E/M. These come to 5952 and 7576 respectively (Carroll rounds them off to 6000 and 7600 respectively). He then calculates the threshold electorates, under each quota, at which a district becomes entitled to one extra seat. He takes it that a district becomes entitled to its first seat when the relevant formula shows its entitlement as exactly 0.5, to a second when its entitlement is exactly to 1.5, and so on. He states that the problem of districts very near a threshold could be solved by enlarging them or reducing them to a size not close to a threshold. His Table 1 (*PPR*, chapter II, § 4) then shows the range of electorates that would qualify a district for any given number of seats. The left-hand column of Table 1 shows the thresholds generated by Carroll's formula $m = (e/Q) - 1$. The right-hand column shows the thresholds generated by the natural formula $m = eM/E$. As with the calculation of the quotas, Carroll rounds these numbers off to the nearest thousand.

Carroll's Table 1 shows two further serious consequences of his switch from a formula based on the natural quota to one based on a constant Droop quota. First, the minimum threshold for getting any seats at all was considerably higher than with

---

36 The source for these figures is F W S. Craig, *British Parliamentary Election Results 1885--1918* (London, Macmillan 1974), especially Table 4.

the natural quota. Second, under Carroll's formula entitlements to seats were not linearly additive, so that entitlements to seats would have risen or fallen as districts were merged or split. Both of these problems may be illustrated by comparing the most thinly and most thickly populated parts of Britain, namely the Scottish Highlands and the East End of London, using the actual figures for 1885 (Figs 3.11 and 3.12).

Constituency	Electors	Votes cast, 1885:		
		Liberal	Crofter	Cons/Ind
Inverness Burghs	3,556	<u>1,709</u>	1,546	
Wick Burghs	2,015	868	<u>913</u>	
Argyll	10,011	670	<u>3,340</u>	2,856
Caithness	4,320	1,218	<u>2,110</u>	
Inverness-shire	9,330	1,897	<u>3,555</u>	2,031
Orkney & Shetland	7,394	<u>3,352</u>		1,940
Ross & Cromarty	10,265	2,925	<u>4,942</u>	
Sutherland	3,185	<u>1,701</u>	1,058	
	50,076	**14,340**	**17,464**	**6,827**

Fig. 3.11 General Election of 1885: the Scottish Highlands
(source for Figs 3.11 and 3.12: Craig, *British Parliamentary Election Results)*
All seats were single-member districts. Winning party underlined

The actual outcome in the Highlands was the election of three Liberals and five Crofters (the Crofters' Party articulated the grievances of those who had been driven off their smallholdings or feared they might be); in the East End, of nine Liberals and five Conservatives. What might have happened with Carroll's scheme?

In the Highlands, he would have been confronted immediately by the impracticality of a minimum threshold of 9000. The Orkney and Shetland Islands, with 7394 electors, could not sensibly have been linked with any other constituency. Apart from them, the Highlands could have been divided either into four single-member districts, or two multi-member districts. Assuming that voting behaviour would have been the same, the outcomes would have been as in Figs 3.13 and 3.14.

Constituency	Electors	Votes cast, 1885:	
		Liberal	Conservative
Bethnal Green NE	7,102	3,095	1,844
Bethnal Green SW	8,265	3,088	2,200
Hackney Central	7,381	2,748	2,941
Hackney N	8,058	2,911	3,327
Hackney S	8,684	3,544	2,602
Shoreditch Haggerston	6,737	2,736	1,259
Shoreditch Hoxton	8,469	3,084	2,047
Tower Hamlets:			
Bow & Bromley	8,887	3,419	2,738
Limehouse	5,954	1,676	2,566
Mile End	5,804	1,442 + 420	2,091
Poplar	9,041	4,090	2,113
St. George	4,317	1,180	1,744
Stepney	6,925	2,045	2,035
Whitechapel	6,140	2,353	1,972
	101,764	**37,831**	**31,479**

Fig. 3.12 General Election of 1885: the East End of London

Constituency	Electors	Seats won
Orkney & Shetland	7,394	1 Lib
Caithness, Sutherland, & Wick	9,520	1 Crofter
Ross & Cromarty	10,265	1 Crofter
Inverness	12,886	1 Crofter
Argyll	10,011	1 Crofter

Fig. 3.13 General Election of 1885: the Scottish Highlands. Hypothetical outcome under Carroll scheme with small electorate districts

Constituency	Electors	Seats won
Orkney & Shetland	7,394	1 Lib
Caithness, Sutherland, Ross &	19,785	1 Crofter
Cromarty & Wick		1 Lib
Inverness & Argyll	22,897	1 Crofter
		1 Lib
		1 Cons

Fig. 3.14 General Election of 1885: the Scottish Highlands. Hypothetical outcome under Carroll scheme with large electorate districts

This simulation shows both sorts of arbitrariness in Carroll's scheme. With large districts, the outcome is much more proportional to votes than with small, as he would expect. But the Crofters' Party would plunge from four seats out of five in the small-district apportionment to two seats out of six in the large-district apportionment; both the Liberals and the Conservatives would gain. With the large-district apportionment, the Liberals would have more seats, in return for fewer votes, than the Crofters.[37] The region would gain a seat because, under Carroll's scheme, the merger of two single-member districts of 12,886 and 10,011 electors produces a three-member district of 22,897 electors.

Figs 3.15 and 3.16 show the results of similar simulations for the East End. Here we have assumed that multi-member districts would have been based on boroughs; that a scheme with small districts would split the largest borough along a line from north-west to south-east; and that a scheme with large districts would amalgamate the three smaller boroughs. Many other schemes would of course have been possible, and might produce equally arbitrary variations in the simulation results

Constituency	N. of seats	Won by Lib.	Won by Cons
Tower Hamlets NE	2	1	1
Tower Hamlets SW	3	2	1
Bethnal Gn & Shoreditch	4	3	1
Hackney	3	2	1
	12	**8**	**4**

Fig. 3.15 General Election of 1885: the East End of London. Hypothetical outcome under Carroll scheme with small electorate districts

37 This result arises from the impracticality of merging Orkney & Shetland with any other constituency.

Constituency	N. of seats	Won by Lib.	Won by Cons
Tower Hamlets	7	4	3
Bethnal Gn, Hackney, & Shoreditch	8	5	3
	15	**9**	**6**

Fig. 3.16 General Election of 1885: the East End of London. Hypothetical outcome under Carroll scheme with large electorate districts

Here again, a change in the size of the districts would change both the total and the party composition of the representation. Comparing the East End with other parts of the country, it would gain by taking advantage of the bias towards the large inherent in Carroll's apportionment scheme; in terms of party advantage, the Conservatives would gain disproportionately from such a move. The ratio of electorates between the Highlands and the East End was approximately 1:2. Their ratio of representation in the actual 1885 apportionment was 8:14. Under a Carroll scheme with the smallest practicable districts it would have been 5:12; with the largest practicable districts, 6:15. Although these variations are less than those permitted by the apportionment system created in 1885 and still with us today, they should certainly have given a meticulous mathematician like Carroll cause for concern.

We have to conclude, then, that Carroll's section on apportionment is a fascinating failure. Fascinating because, typically, he pursues the logic of his argument to its conclusion; but a failure because he does not consider its perverse consequences.]

Part 4. Reprints of original material

4.1. Lewis Carroll: The Principles of Parliamentary Representation, with Supplement and Postscript to Supplement

THE PRINCIPLES OF PARLIAMENTARY REPRESENTATION

PREFACE

Through all the dust and din of the present controversy, four things, at least, are surely clear to all thinking men:-

First, that it would be an unmitigated evil to have a General Election with the new Franchise, but without a new Distribution of Seats;

Secondly, that there would be no difficulty in avoiding all risk of such a catastrophe, *PROVIDED THAT* a clause were added to the Franchise-Bill, enacting that it "shall not be put into operation until a Redistribution-Bill has also been passed";

Thirdly, that there would be no difficulty in both parties agreeing to such a clause, *PROVIDED THAT* each felt secure against the other party obtaining an unfair advantage in the Redistribution;

Fourthly, that there would be no difficulty in making this secure, *PROVIDED THAT* some general principles, making it impossible for either side to obtain any such advantage, could be discovered and accepted by both parties.

It is in the profound conviction that such principles exist, and that they can be as clearly formulated, and as fully proved, as the principles of any other Science, that I venture to address these pages to all interested in the matter.

C. L. D.

Ch. Ch., Oxford,
Nov. 5, 1884

CHAPTER I.

Desiderata.

The chief *desiderata* seem to be as follows:-

(1) That each Elector should have the same chance of being represented in the House. (Under *any* system, *some* Electors must be left unrepresented.)

(2) That each Elector, who is represented at all, should be represented by the same fraction of a Member. Or (which is the same thing) that each Member should represent the same number of Electors. Or (which is the same thing) that the number of Electors, needed to secure the return of a Member, should be uniform throughout the Kingdom.

(3) That the number of unrepresented Electors should be as small as possible.

(4) That the proportions of political parties in the House should be, as nearly as possible, the same as in the whole body of Electors.

(5) That the process of voting should be as simple as possible.

(6) That the process of counting the votes, and announcing the result, should be as simple as possible.

(7) That the waste of votes, caused by more votes being given for a Candidate than are needed for his return, should be as far as possible prevented.

(8) That the result of a local Election should depend as much as possible on the wishes of the Electors in that District, and as little as possible on chance.

(9) That the Electors in a District should be, as far as possible, uninfluenced by the results of Elections in other Districts.

CHAPTER II.

Principles to be observed in forming electoral Districts, and in determining, for each District, how many Members it shall return.

§ 1. *Number of Members in House.*

There seems to be no sufficient reason, *a priori*, for any change in this particular. It would probably be best to take 660 as the number to be generally aimed at, though holding ourselves free to modify this as circumstances might require.

§ 2. *Number of electoral Districts; whether to be equal or unequal; &c.*

The two extreme cases are (1) to have as many Districts as Members, each to return one Member, in which case the Districts should of course be equal; (2) to form the whole Kingdom into one District.

In the first case (a method that has been much advocated) it is only a bare majority in each District who are represented. For it must not be supposed that all who vote for a Member are duly represented by him. If a District contains 20,001

Electors, so that 10,001 are enough to return a Member, all additional votes are absolutely wasted: hence only 10,001 Electors in that District are represented in Parliament; the other 10,000, whether they vote for the successful Candidate, or for a rival, or even if there be no contest at all, are unrepresented. This method, then, leaves nearly half the whole body of Electors unrepresented.

The injustice of this method may be illustrated from two points of view. Suppose a bare majority of the Electors to be of one party, and the rest of the opposite party; e.g. let 6/11ths be 'red' and 5/11ths 'blue'. Then, as a matter of abstract justice, about 6/11ths of the House ought to be 'red', and 5/11ths 'blue'. But practically this would have no chance of occurring: if the 'reds and 'blues' were evenly distributed through the Kingdom, a 'red' would be returned in every District, and the whole House would be of one party! Yet this distribution is, by the Laws of Probability, more likely than any other one distribution, and, the nearer the distribution to the most probable one, the nearer we come to this monstrous injustice.

The other way of looking at it is almost as telling. Suppose the House to have been elected, and that 6/11ths of the Members are 'red', and 5/11ths 'blue': all we could learn from this, as to the views of the Electors, would be that 6/22ths [sic] (about 28%) are 'red', and 5/22ths (about 23%) 'blue': as to the other 49%, we should know absolutely nothing - if they were all 'red' (i.e. if 3/4ths of the Electors were 'red'), or all 'blue' (i.e. 7/10ths of the Electors 'blue'), it would make no difference in the House.

Taking this first extreme, then, as yielding the *maximum* of injustice which can be effected by arrangement of Districts, and observing that, if each District returned 2 Members, only 1/3rd of the Electors (on the assumption that each Elector has only one vote - an arrangement whose justice we shall hereafter prove) would be unrepresented, if 3 Members, only 1/4th, and so on, we see that the fewer and larger the Districts, i.e. the greater the number of Members which, on an average, each District returns, the fairer the result: till we come to the other extreme, where the whole Kingdom is formed into one District returning 660 Members, in which case only 1/661th [sic] of the whole body of Electors would be left unrepresented. A general Election, with so gigantic a District, would of course be impracticable: and probably Districts, returning 6 Members each, would be about as large as could be conveniently dealt with: but very small Districts should be, as far as possible, avoided.

I find, in the *Standard* for October 10, 1884, a very good instance of the injustice done by sub-dividing large electoral Districts. "The Birmingham Conservatives are, a Correspondent telegraphs, keenly discussing the Government Redistribution Scheme. The clause which apportions 6 Members to Birmingham gives much dissatisfaction in Conservative circles. It is contended that, if the borough is to be divided into three electoral Districts, each District to have 2 Members, the Liberals could so manipulate the voters as to be certain of returning the whole of the 6 Members". Now, assuming that each Elector is to have one vote only, the Liberals could only do this by mustering more than two-thirds of the votes in each District; i.e. they must be 67%, or more, or the whole body of Electors in

Birmingham. But, if the three Districts were made one, it would need about one-seventh of the whole (i.e. 14 and 2/7ths %) to return one Member. Hence 67% could only return 4 of the 6 Members: it would require 71% to return as many as 5; and they could not return all 6, unless they were 86% of the whole body.

Taking it as proved, then, that single-Member Districts should be in all cases avoided, and that all such should be grouped together, so as to form Districts returning at least 2 Members each, and, wherever it is possible, 4 or 5 or even more, we need only add, as a general remark, that, the more we equalise the Districts, the more we equalise the chance that each Elector has of being one of those represented in the House. Thus, in a District, returning 2 Members, the chance is 2/3rds; with 3 Members, it is 3/4ths; and so on.

§ 3. *Formula for determining, for each District, how many Members it shall return.*

A preliminary question must here be asked, viz. are we to count population, or Electors only? I do not think it matters much which, as they probably vary nearly together, i.e. a District having twice the population of another would probably have twice as many Electors. The Formula can best be determined for the number of *Electors*: but if, in using it, the number of population be substituted, it will make no important difference in the result.

The formula will of course have to be modified for each case, if it be agreed to give political weight to differences in rateable property, or to the distinction between town and country voters: and for this purpose rules would have to be laid down.

Now, taking 'e' to represent, for any one District, the number of Electors, and 'm' the number of Members to be assigned to that District, and assuming that each Elector has only one vote, we require a formula giving m in terms of e. This formula must evidently be such as will secure that every Member in the House shall, as far as possible, represent the same number of Electors.

Now, whatever be the quota of recorded votes, which is necessary and sufficient, *before the poll is closed*, to make it certain that '*A*' will be returned, that is the number of Electors whom *A* will represent in the House. He cannot represent *less*, for this number is *necessary*; and he cannot represent *more*, for it is *sufficient*, so that all additional votes are superfluous. Let us call this necessary and sufficient quota 'Q'.

Now, in order that Q may be *sufficient*, it must not be possible for m other Candidates to obtain Q votes each; i.e. $(m + 1).Q$ must be greater than e; i.e. Q must be greater than $\frac{e}{m+1}$. Also, in order that Q may be *necessary*, it must be the whole number *next* greater than this fraction. Hence, approximately, $Q = \frac{e}{m+1}$; i.e.

$$m = \frac{e}{Q} - 1$$

This, then, is the formula required. An example will make it clear. Suppose the universal quota to be 6,000: then a District containing 50,000 Electors would have 7 Members assigned to it.

We have yet to find a formula for determining Q. Let 'e_1' be the number of Electors in district No. 1, 'e_2' the number in No. 2, and so on; let 'm_1' be the number of Members assigned to District No. 1, 'm_2' the number assigned to No. 2, and so on; also let 'E' be the total number of Electors in the Kingdom, 'M' the number of Members in the House, and 'D' the number of Districts. Then we have

$$(m_1 + 1).Q = e_1$$
$$(m_2 + 1).Q = e_2$$
&c.

$$\therefore\ (M + D).Q = E; \qquad \text{i.e. } Q = \frac{E}{M + D};$$

$$\therefore\ m = e.\frac{M + D}{E} - 1.$$

§ 4. *Tables calculated by the preceding Formulæ.*

Let us suppose the 2,000,000 new Electors to be already enfranchised, thus making the total Electorate about 5,000,000. Let us further assume the number of electoral Districts to be 180, so that each will return, on an average, 3 and 2/3rds of a Member.

Let M = No. of Members in House = 660.
D = No. of Districts = 180.
e = No. of Electors in a District.
E = total No. of Electors = 5,000,000.
p = population in a District.
P = total population = 36,000,000.
Q = universal quota, to be aimed at.
m = No. of Members assigned to a District.

Then $\dfrac{E}{M + D} = \dfrac{5,000,000}{840}$ = about 6,000;

$$\therefore\ m = \frac{e}{6,000} - 1 \qquad \text{.................................... (a)}$$

It will be worth while to contrast with this the 'rough and ready' method of assigning Members in proportion to the number of Electors, so that m:e :: M:E. This gives us

$$m = e.\frac{M}{E} = e.\frac{660}{5,000,000} = \frac{e}{7,600} \qquad \text{.......... (b)}$$

In the following Table [Table I], the second column gives the number of Members to be returned by a District, the first the number of Electors by Formula (a), and the third the same by Formula (b).

TABLE I.

e, by (a)	*m*	*e*, by (b)
9,000		4,000
	1	
15,000		11,000
	2	
21,000		19,000
	3	
27,000		27,000
	4	
33,000		34,000
	5	
39,000		42,000
	6	
45,000		49,000
	7	
51,000		57,000
	8	
57,000		65,000
	9	
63,000		72,000
	10	
69,000		80,000

The numbers, in the first and third columns, have been calculated by giving to *m*, in the preceding Formulæ, the successive values one-half, 3-halves, 5-halves, &c. Hence we see that, by Formula (a), a District containing between 9,000 and 15,000 Electors must have between one-half and 3-halves of a Member (i.e. must have *one* Member) assigned to it: and so on. If a District contained almost exactly 15,000, it could not fairly be determined, by this Table, whether it ought to return one Member, or two. In such a case, it would be best to change the boundaries of the District, so as to increase or diminish the number of Electors by 2,000 or so.

Comparing the results of the two Formulæ, we see that, for Districts whose population is about 27,000, it matters very little which Formula we use: but, for small Districts, Formula (b) assigns too many Members, and, for large Districts, too few; e.g. 13,000 Electors ought to return only one Member - Formula (b) gives them two; 60,000 ought to return 9 - formula (b) gives them 8.

We will now examine the effect of counting the population of a District, and not the Electors only [Table II].

Here, for $\frac{E}{M+D}$, we must substitute $\frac{P}{M+D}$;

i.e. $\frac{36,000,000}{840}$, i.e. about 43,000.

Hence Formula (a) becomes

$$m = \frac{e}{43,000} - 1 \quad \ldots\ldots\ldots\ldots\ldots\ldots\ldots\ldots\ldots\ldots \quad (c)$$

Also Formula (b) becomes

$$m = e.\frac{660}{36,000,000} = \frac{e}{54,500} \quad \ldots\ldots\ldots\ldots\ldots \quad (d)$$

TABLE II.

e, by (c)	*m*	*e*, by (d)
64,000		27,500
	1	
107,000		82,000
	2	
150,000		136,500
	3	
193,000		191,000
	4	
236,000		245,500
	5	
279,000		300,000
	6	
322,000		354,500
	7	
365,000		409,000
	8	
408,000		463,500
	9	
451,000		518,000
	10	
494,000		572,500

Comparing this with Table I, we see that, provided only it be true that the number of Electors in a District is always about 5/36ths of the population, the substitution of number of population for number of Electors will suffice for all practical purposes; and, seeing that there is evidently a tendency to go by population, and that it is much more easy to take the population of a District than to estimate what will be the number of its Elector when the Franchise-Bill is passed, the first column of Table II. is probably the best to employ.

CHAPTER III.

Principles to be observed in conducting Elections.

§ 1. *Number of Votes each Elector may give.*

The two extreme cases are (1) to let each Elector give as many votes as there are Members to be returned by the District; (2) to let him give one vote only.

The effect of each of these methods, and of the intermediate methods which lie between them, will be best understood by considering the following Tables of percentages.

We will first find the general formulæ for determining what number of Electors, in a given District, is necessary and sufficient to secure the return of one Candidate, of 2, of 3, &c..

Let e = No. of Electors in the District,
m = ... Members assigned to it,
v = ... votes each Elector can give,
s = ... seats it is desired to fill,
x = ... Electors required.

Also let it be assumed that an Elector may not give 2 votes to the same Candidate. (N.B. 'cumulative' voting is discussed at p. 162).

Now, in order that x may be *sufficient* to fill s seats, it must be large enough to make it impossible for the other $(e - x)$ Electors to fill $(m + 1 - s)$ seats; since the two events are incompatible, so that, if the latter were possible, the former would be impossible. To effect this, each of the s Candidates must have more votes than it is possible to give to each of $(m + 1 - s)$ rival Candidates.

In order that x may be *necessary*, it must be only *just* large enough for the purpose.

It will be necessary to consider the following 4 cases separately. Observe that > means 'greater than', ≤ means 'not greater than', and ∴ means 'therefore'.

Case (a) v is $\leq s$, and also $\leq (m + 1 - s)$;
Case (b) ..... $> s$, but $\leq (m + 1 - s)$;
Case (c) ..... $\leq s$, but $> (m + 1 - s)$;
Case (d) ..... $> s$, and also $> (m + 1 - s)$.

In case (a), the x Electors can give vx votes, which, divided among s Candidates, supply them with $\frac{vx}{s}$ votes apiece. Similarly, the $(e - x)$ Electors can give $v.(e - x)$ votes, which, divided among $(m + 1 - s)$ Candidates, supply them with $\frac{v.(e-x)}{m+1-s}$ votes apiece. Hence we must have

$$\frac{vx}{s} > \frac{v.(e-x)}{m+1-s},$$

where v divides out;

$$\therefore x.(m + 1 - s) > se - sx;$$

$$\therefore\ x.(m+1) > se;$$

$$\therefore\ x > \frac{se}{m+1}.$$

In case (b), each of the x Electors can only use s of his v votes, since he can only give *one* to each Candidate: hence the x Electors can only give sx votes, thus supplying s Candidates with x votes apiece. But the $(e - x)$ Electors can, as in case (a), supply $(m + 1 - s)$ Candidates with $\frac{v.(e-x)}{m+1-s}$ votes apiece. Hence we must have

$$x > \frac{v.(e-x)}{m+1-s};$$

$$\therefore\ x.(m+1-s) > ve - vx;$$

$$\therefore\ x.(m+1-s+v) > ve;$$

$$x > \frac{ve}{m+1-s+v}.$$

In case (c), the x Electors can, as in case (a), supply s Candidates with $\frac{vx}{s}$ votes apiece. But each of the $(e - x)$ electors can only use $(m + 1 - s)$ of his votes: hence the $(e - x)$ Electors can only give $(m + 1 - s).(e - x)$ votes, thus supplying $(m+ 1 - s)$ Candidates with $(e - x)$ votes apiece. Hence we must have

$$\frac{vx}{s} > e - x;$$

$$\therefore\ vx > se - sx;$$

$$\therefore\ x.(s+v) > se;$$

$$x > \frac{se}{s+v}.$$

In case (d), the x Electors can, as in case (b), supply s Candidates with x votes apiece. And the $(e - x)$ Electors can, as in case (c), supply $(m + 1 - s)$ Candidates with $(e - x)$ votes apiece. Hence we must have

$$x > e - x;$$

$$\therefore\ 2x > e;$$

$$x > \frac{e}{2}.$$

Tabulating these results we have the following formulæ.

	Data	*Formulæ*
(a)	$v \leq s$ $\leq m+1-s$	$x > \frac{se}{m+1}$
(b)	$v > s$ $\leq m+1-s$	$x > \frac{ve}{m+1-s+v}$
(c)	$v \leq s$ $> m+1-s$	$x > \frac{se}{s+v}$
(d)	$v > s$ $> m+1-s$	$x > \frac{e}{2}$

By these formulæ the following Table [Table III] is calculated. It shows, for a given District, what percentage of the electors is necessary and sufficient to secure the return of *one* Candidate, of 2, of 3, &c..

The 2nd line in the 3rd section represents the well-known "three-cornered constituency". Observe (by comparing it with the next line) that it makes it too hard for a minority to fill *one* seat, and too easy for a majority to fill *all.*

TABLE III.

No. of Members returned. by District.	No. of votes each Elector can give.	No. of Seats it is desired to fill.					
		1	2	3	4	5	6
1	1	51					
2	2	51	51				
	1	34	67				
3	3	51	51	51			
	2	41	51	61			
	1	26	51	76			
4	4	51	51	51	51		
	3	43	51	51	58		
	2	34	41	61	67		
	1	21	41	61	81		
5	5	51	51	51	51	51	
	4	45	51	51	51	56	
	3	38	43	51	58	63	
	2	29	34	51	67	72	
	1	17	34	51	67	84	
6	6	51	51	51	51	51	51
	5	46	51	51	51	51	55
	4	41	45	51	51	56	61
	3	34	38	43	58	63	67
	2	26	29	43	58	72	76
	1	15	29	43	58	72	86

In examining this Table, we notice, first, the uniformity of the *upper* line in each section (i.e. the percentages required when each Elector can give as many votes as there are seats to fill). Here, in every case, more than half the Electors must agree, in order to fill one single seat: but, when once this number have mustered, they have it in their power to fill *all* the seats! *'C'est le premier pas qui coûte'*.

This absurdity diminishes gradually, from line to line, as we look down each section; the lowest line (i.e. the percentages required when each Elector can give one vote only) being always the most reasonable. One of the most startling anomalies is the 4th line of the 6th section. Here we see that, out of 100 Electors, we must muster 34 in order to fill *one* seat: with four more Electors, we can fill the second seat: with five more, the third: but 'then comes the tug of war'; to win the fourth seat, we actually need *fifteen* more Electors!

Lastly, comparing together the lowest lines of the several sections, we notice that they gradually improve as we move down from section to section, requiring a smaller percentage to fill *one* seat, thus giving a minority a better chance of being represented, and a larger percentage to fill *all*, thus leaving a smaller number unrepresented. This last figure (the right-hand end of each lowest row) represents the percentage of the Electors in the Kingdom who would be represented in the House, supposing all the Districts similar to the one under consideration: and this percentage we find to rise, from 51 in the case of single-Member Districts, to 86 in the case of six-Member Districts.

The obvious conclusion is - let the Districts be as *large* as possible, and let each Elector give *one* vote only.

The effect, on the composition of the House, will be yet more clearly seen by considering the following three Tables, which are calculated on the assumption that, in any District, all proportions, between 'red' and 'blue', are equally probable, and that 6/11ths of the House are 'red' and 5/11ths 'blue'. Table IV. gives the percentage of the whole body of Electors represented by the 'red' Members, Table V. the percentage represented by the 'blue', and Table VI. the percentage unrepresented:-

TABLE IV.

		Number of votes each Elector can give					
		6.	5.	4.	3.	2.	1.
Number of	1.						28
Members	2.					28	37
assigned	3.				28	36	42
to each	4.			28	35	40	44
District	5.		28	33	39	43	46
	6.	28	32	36	40	44	48

TABLE V.

		Number of votes each Elector can give					
		6.	5.	4.	3.	2.	1.
Number of	1.						23
Members	2.					23	31
assigned	3.				23	30	34
to each	4.			23	29	34	37
District	5.		23	28	32	36	38
	6.	23	27	30	34	37	38

TABLE VI.

		Number of votes each Elector can give					
		6.	5.	4.	3.	2.	1.
Number of	1.						49
Members	2.					49	32
assigned	3.				49	34	24
to each	4.			49	36	26	19
District	5.		49	39	29	21	16
	6.	49	41	34	26	19	14

By inspecting these Tables, we see two things:-

First, that the fewer and larger the Districts, i.e. the greater the number of Members returned (on an average) by each District, the more equitable the result. this conclusion we have already arrived at, from general considerations. (See p. 153). We observe, further, that the advantage, in fairness of result, increases rapidly at first and more slowly afterwards. For instance, in Table VI, if each Elector be allowed one vote only, the change from single-Member to two-Member Districts changes the percentage of unrepresented Electors from 49 to 32 (i.e. deducts about 1/3rd); whereas the change, from 5-Member to 6-Member Districts, only changes the percentage from 16 to 14 (i.e. deducts only 1/8th). The conclusion is that *the* important point is to have as few single-Member, and even as few 2-Member, districts as possible; but that, when we have got as far as to districts returning 4 or 5 Members each, it is hardly worth while to go further.

Secondly, we see that the fewer the number of votes (down to the least possible, viz. '*one*') that each Elector is allowed to give, the more equitable the result. We observe, further, that the advantage, in fairness of result, increases slowly at first and more rapidly afterwards. For instance, in Table VI, if 6 Members be assigned to a District, the change from 6 votes to 5 only changes the percentage of unrepresented Electors from 49 to 41 (i.e. deducts less than 1/6th); whereas the change from 2 votes to one changes it from 19 to 14 (i.e. deducts more than 1/4th). We observe, further, that the system of allowing each Elector as many votes as there are seats to fill produces, in *every* case, the same result, (the most inequitable that it is possible to produce by any variation in these data), viz. that it leaves about 49% of the Electors unrepresented. The system (already discussed at p. 153) of "equal electoral Districts, each returning one Member" is only a particular instance of this general law.

The method of 'cumulative voting' (where an Elector can give two or more votes to the same Candidate) will usually have no other effect than to increase the 'specific gravity' - so to speak - of a vote. Let each Elector have 4 votes, with permission to 'lump' them if he chooses, and in the end you will find most of the votes given in lumps of 4, and the result much the same as if each Elector had had *one* vote only.

The conclusion is that *the* important point is to let each Elector give *one* vote only.

§ 2. *Formula for determining, after the poll is closed, the quota of Votes needed to return a Member.*

By a process, exactly similar to that employed at p. 154, we may prove that, it 'r' be the number of recorded votes, and 'm' the number of Members to be returned, the quota must be just greater than $\frac{r}{m+1}$. For example, if 55,000 votes had been given, and the District had to return 6 Members, the quota needed to return one Member would be just greater than 7,857 and 1/7th: i.e., a Member, having 7,858 votes, would be returned. Similarly, anything just greater than 15,714 and 2/7ths would be enough (if the votes could be reckoned *en masse*) to return 2 Members: i.e., if 2 Members of the same party had 15,715 votes between them, both could be returned. We shall prove, further on, that such reckoning of votes is equitable and ought to be provided for.

This quota must be carefully distinguished from the one discussed at p. 154. If a District, returning one Member, contains 10,001 Electors, the quota needed, *before the poll is closed*, to make it certain that 'A' will be returned, is 5,001; but, if only 8,001 vote, the quota needed, *after the poll is closed*, to return him, is only 4,001. For the purpose of *assigning Members to a District*, it is fair to proceed as if *all* the Electors were sure to vote; but, for the purpose of *returning Members*, we can count only the votes that are actually recorded.

§ 3. *Method for preventing waste of Votes.*

Assuming it to be agreed that each District is to return 2 or more Members, and that each Elector is to give one vote only, we have now to consider what is to be done when 2 or more Candidates of the same party have got, among them, enough votes to be returned, but when some have got more than the quota, and others less. It is obviously not fair that the party should fail in bringing in their rightful number of Members, merely by an accidental disarrangement of votes; but how to make an equitable transfer of the superfluous votes is by no means so obvious.

Various methods have been proposed for this: of which I will consider two:-

(1) "The Proportional Representation Society" proposes to let each Elector hand in a list of Candidates, marked in the order of his preference; and that his vote, if not required for his No. 1, should be transferred to his No. 2, and, if not required for him, then to No. 3, and so on. One great objection to this method is the confusion it would cause in the mind of an ignorant Elector, who, though quite able to name his favourite Candidate, would be utterly puzzled if told to arrange 5 or 6 names in order of merit. But a much stronger objection is the difficulty of deciding to *which* of the remaining Candidates the surplus votes shall go: e.g. if 8,000 be the quota needed to return a Member, and if 6,000 lists be headed 'AB', and 4,000 'AC', *which* 2,000 are to be transferred? Mr. J. Parker Smith, in a Pamphlet entitled "Preferential Voting", says (at p. 2), "The course which is exactly fair to B and C is that the votes which are

transferred should be divided between them in the same proportion as that in which the opinions of the whole number of *A*'s supporters is divided". (This would require, in the above instance, that 3/5ths of the 2,000, i.e. 1,200, should be taken from the '*AB*' lists, and 2/5ths, i.e. 800, from the '*AC*' lists). He adds, "This principle avoids all uncertainty, and is indisputably fair". He then proceeds to show that if, instead of counting and arranging the surplus votes, they be taken "in a random order", the chances are very great that they will come out nearly in this proportion. And he further adds (at p. 4), that "the element of chance will not be of importance as between the different parties, but only as between different individual Candidates of the same party". Now all this rests on the assertion that this mode of dividing the surplus votes, whether effected by counting or left to chance, is "indisputably fair": and this assertion I entirely deny. The following instance will serve the two purposes, of showing that this method may easily lead to gross injustice, and of showing that the difficulty may easily arise between candidates of opposite parties.

Take a town of 39,999 Electors, returning 3 Members, so that 10,000 votes will suffice to return a Member; let there be 4 'red' Candidates, *A*, *B*, *C*, *D*, and one 'blue', *Z*; and let there be 21,840 lists "*ABD*", 10,160 "*ACB*", and 7,999 "*Z*". There can be no shadow of doubt that, as a matter of justice, *A*, *B*, *C* ought to be returned, since there are more than two full quotas who put '*AB*' first, and, over and above these, more than one quota who put '*AC*' first. Let us see what, under the Society's present rules, would be the most probable result.

The 32,000 lists headed "*A*" are of two kinds, bearing to each other the ratios of the numbers 273, 127. Hence the certain event, if the lists are divided by rule, and the most probable event, if they are divided at random, is that the 10,000 lists, used in returning *A*, will contain 6,825 "*ABD*" and 3,175 "*ACB*". Erasing "*A*" from the remaining lists, we have now in hand 15,015 "*BD*", 6,985 "*CB*" and 7,999 "*Z*"; so that *B* is returned.[1] Erasing "*B*" from the remaining lists, we now have 5,015 "*D*" 6,985 "*C*", and 7,999 "*Z*"; so that *Z* is returned with a majority of more than 1,000 over *C*. And the 'reds' must derive what consolation they can from the reflection that their rejected Candidate really had 2,161 more supporters than the successful 'blue'!

While fully agreeing, then, with the Proportional Representation Society as to the propriety of allowing only one vote to each Elector, I think I have sufficiently proved the fallacy of its method for disposing of surplus votes.

(2) A mechanical method of recording votes was suggested, in a letter signed "F. R. C.", in the *St. James' Gazette* for Aug. 1. Each Elector is to pass (unseen) through one of a set of turnstiles, (each Candidate having a separate turnstile), which will mechanically record his vote. The records are to be periodically examined, and the results placarded outside, in order that Electors, on seeing that a Candidate has already got votes enough to secure his return, may cease to vote for him. Several objections, each by itself fatal, may be made to this method. One is that, if the

---

1 [Black annotated one of his copies with 'This does not in fact follow from Hare's scheme'.]

periods were short enough to prevent waste of votes, the inspection would destroy the secrecy of the ballot, as it would be known who had just voted, and the result of his voting would be at once placarded; whereas, if the periods were long enough to avoid this, time would be allowed for large waste of votes. Another is that, as the quota, necessary to return a Candidate, could not be fixed till the poll had closed, it would be impossible to know, during the Election, whether a Candidate had or had not received votes enough to secure his return. Another is that, if part of the machinery went wrong, so as (for instance) to record a total of votes greater than the number of Electors, the mistake could not (as it can with voting-papers) be rectified, but the Election would have to be held over again.

Having proved, then, that the method of arranged lists will not serve fairly to dispose of surplus votes, and yet that we cannot prevent such votes being given, we have now to find, if possible, a fair method for disposing of them. Clearly *somebody* must have authority to dispose of them: it cannot be the Elector (as we have proved); it will never do to refer it to a Committee. There remains *the Candidate himself, for whom the votes have been given.* This seems to solve the whole difficulty. The Elector must understand that, in giving his vote to *A*, he gives it him as his absolute property, to use for himself, or to transfer to other Candidates, or to leave unused. If he cannot trust the man, for whom he votes, so far as to believe that he will use the vote for the best, how comes it that he can trust him so far as to wish to return him as a Member?

§ 4. *Method for preventing the Electors in one District from being influenced by the results of Elections in other Districts.*

That Electors are liable to such influences may be proved both *a priori* and *a posteriori*. On the one hand, it is a tendency of human nature, too well-known to need proving, to surrender one's own judgment in order to be on the winning side. In the words of the immortal Mr. Pickwick, "it's always best on these occasions to do what the mob do". "But suppose there are two mobs?" suggested Mr. Snodgrass. "Shout with the largest", replied Mr. Pickwick. On the other hand, no one, who has ever watched the progress of a General Election, can need to be reminded how obviously the local Elections of the later days have 'followed suit', under the irresistible influence of those of the earlier days. "The secret of success", it has been well said, "is to succeed": and there can be little doubt that the party, which fails in carrying a majority of the local Elections at first, is heavily handicapped during the rest of the contest.

Supposing it admitted that such an influence does exist in General Elections as now managed, and that it is an influence to be avoided, the remedy is not far to seek: let the local Elections be so arranged that all, or nearly all, the results may be announced at the same time.

This arrangement would no doubt be unwelcome to certain 'pluralists', who are now able to vote in several different Districts. Possibly, in such exceptional cases, voting-papers might be allowed. But, even if no remedy could be found, the justice of allowing one Elector to vote as if he were, "like Cerberus, three gentlemen at once", seems so doubtful that the objection hardly deserves serious consideration.

§ 5. *Conduct of Elections.*

The practical working of the principles, which have now been demonstrated, would be as follows:- When the poll is closed, let the total number of votes recorded be divided by the number of Members to be returned increased by one, and let the returning-officer announce the whole number next greater than the quotient as the quota needed to return *one* Member. Similarly, the whole number next greater than twice the quotient will be the quota needed to return *two*, and so on.

Let him further announce the number of votes given for each Candidate, and also announce as "returned" any Candidate who has received the quota needed to return *one.* If there are still Members to return, let him appoint a time and place for all the Candidates to appear before him; and any two or more Candidates may then formally signify that they wish their votes to be clubbed together, and may nominate so many of themselves as can be returned by the votes so clubbed. They must of course include in their nomination any of themselves who have been already declared to be returned. Let the returning-officer add together the votes of these Candidates, and, if the amount be not less than the necessary quota, let him declare to be duly returned the Candidates so nominated.

As an example, suppose that a District is to return 5 Members, and that there are 4 'red' Candidates, *A*, *B*, *C*, *D*, and 3 'blue', *X*, *Y*, *Z*. Then the returning-officer might announce as follows:-

Votes given for

C	15,000
X	9,000
D	8,001
Z	8,000
B	7,500
A	6,500
Y	6,000
	6|60,001
	10,000 and 1/6th.

Quota needed to return

1 Member	10,001
2 Members	20,001
3 Members	30,001
4 Members	40,001
5 Members	50,001

I hereby declare C to be duly returned.
Four vacancies remain to be filled.

(Signed)

____

The Candidates might then appear before the returning-officer, and *B*, C, *D* might formally declare that they wished to club their votes; and, as the sum total of their votes is 30,501, they would be declared to be "returned": similarly, *X*, *Y*, *Z* might club their votes, naming *X* and *Z* as the Candidates to be returned; and, as the sum total of their votes is 23,000, *X* and *Z* would be declared to be "returned".

Such Candidates would have to sign some such paper as the following:-

We, the undersigned, for whom the recorded votes, as stated below, amount to ___________, which is not less than ___________, the quota announced as needed to return ________ Candidates, hereby declare that we desire the said votes to be clubbed together. And we nominate, as Candidates whom we desire to be returned by the said votes, in addition to ____, who have been already declared to be duly returned, _____________________.

Signed,

Names	*Votes*
Sum total of votes	

This method would enable each of the parties in a district to return as many Members as it could muster the proper quota for, no matter how the votes were distributed. There would be no risk of a seat being left vacant through rivalry between two Candidates of the same party: an unwritten law would soon come to be recognised - that the one with fewest votes should give way. With Candidates of two opposite parties, such a difficulty could not arise at all: one or other of them could always be returned by the surplus votes of his own party. the only exception to this would be the occurrence (a very rare one) of an exact balance of votes. this might happen, even in the case of a single-Member constituency, if each of 2 Candidates got exactly half the votes. Of course, in such a case, somebody must give a casting-vote.

CHAPTER IV.

Final Summary.

The main points, which I claim to have made good in this little treatise, are as follows:-

(1) That electoral Districts should be so large as to return, on an average, 3 or more Members each: and that single-Member Districts should be, as far as possible, done away with.

(2) That Members should be assigned to the several Districts in such numbers that the quota, needed to return a Member, should be tolerably uniform throughout the Kingdom.

(3) That each Elector should give one vote only.

(4) That all votes given should be at the absolute disposal of the Candidate for whom they are given, whether to use for himself, or to transfer to other Candidates, or to leave unused.

(5) That the Elections in the several Districts should terminate, as nearly as possible, at the same time.

As a practical conclusion to this treatise, I venture to suggest the following ideal Schedule of General Resolutions, such as might fairly be agreed on by all parties, and thus tend to the peaceful termination of this deplorable controversy.
(N.B. The *numbers* here suggested are merely tentative, and capable of being modified *ad libitum*.)

General Resolutions

1. The House shall consist of 660 Members.
2. There shall be 180 electoral Districts.
3. No District shall contain less than a population of 60,000, or more than 500,000.
4. A District, whose population is between 60,000 and 105,000, shall have one Member assigned to it; between 105,000 and 150,000, two Members; and so on, in accordance with the following Table:-

Population.	Members.
60,000	
	1
105,000	
	2
150,000	
	3
195,000	
	4
240,000	
	5
280,000	
	6
320,000	
	7
365,000	
	8
410,000	
	9
455,000	
	10
500,000	

5. If the population of a District be very near to one of the above-named numbers, its boundaries shall be altered so as to increase, or diminish, the population, by not less than 10,000.

6. If it be agreed to give political weight to differences in rateable property, or to the difference between town and country votes, this shall be done by modifying the number of Members assigned by the above Table.

7. The procedure at a local Election shall be as follows:- Each Elector shall give one vote only. When the poll is closed, the number of recorded votes shall be divided by the number of Members to be returned increased by one, and the returning-officer shall announce the whole number, next greater than the quotient, as the quota needed to return one Member; the whole number, next greater than twice the quotient, as the quota needed to return two Members; and so on. He shall also announce the number of votes recorded for each Candidate, and shall declare to be duly returned any Candidate who has obtained the quota. If any vacancies remain to be filled, he shall appoint a time when the Candidates shall appear before him, and any two or more of them may then formally signify their desire to club their votes, and may nominate, as Candidates to be returned by those votes, so many of themselves as the votes suffice for: provided always that they include, in such nomination, any of themselves who have been already declared to be returned. And, if the sum total of the votes so clubbed be not less than the quota needed to return the Candidates so nominated, the returning-officer shall declare to be duly returned all of them who have not been already so declared.

8. The local Elections shall be so arranged that their results may be announced, as nearly as possible, at the same time.

4.1.2. The Principles of Parliamentary Representation. Supplement.

Now that the public are beginning to realise the justice of the demand for "proportional representation", the day cannot be far off when they will understand the gross injustice (so clearly proved by Lord Salisbury in the *National Review* for October)[2] of the "single-member" districts, and will recognise as true the main contention of the Society for Proportional Representation, that each district should return several Members, and that each elector should have *one* vote only.

May I, as one who has given much thought to this subject, point out a serious mistake which the Society is making in the details of its method, and which is in my belief greatly damaging its cause?

It is fairly obvious, and by this time pretty generally known, that, to obtain the quota of votes necessary and sufficient to return a candidate, when each elector has one vote only, we must divide the total number of votes by the number of

---

2 [3rd Marquess of Salisbury, 'The value of redistribution: a note on electoral statistics', *National Review* 4 (1884), pp. 145--62.]

members to be returned plus one, and that the whole number, next above the quotient, is the required quota: e.g. if the district returns 3 members, and 4000 votes are recorded, we divide by 4, and the quota is 1001.

It is also obvious that it will often happen to a candidate to poll more votes than he needs, and the question arises, how are the spare votes to be utilised? The answer of the Society is "Let the voter mark on his paper his second-best man, his third-best, and so on: and, if his first man is already returned, let his vote be used for his second, and so on". This method is complicated and likely to puzzle voters: but such an objection might well be set aside, if only it were just.

But a second question arises. Suppose that *A* has votes to spare, and that some of the papers, headed *A*, have *B* as second-man, while others have *C*: which papers are we to transfer? To this the Society replies "The absolutely just method would be to count how many papers have *B* as second-man, and how many have *C*, and to maintain this proportion in the transferred votes: e.g. if $^3/_4$ths of the whole set of papers, headed *A*, have *B* as second-man, then *B* ought to have $^3/_4$ths of the transferred votes. But, if the votes are simply shuffled and drawn at random, the probability is that this proportion will be almost exactly secured: so that the above calculation may be dispensed with".

Thus the Society is in effect making two assertions, both of which are demonstrably incorrect: one, that it is only when an elector's first-man is already returned that his vote would be used for his second-man; the other, that the Society's method for transferring spare votes would always secure a just result.

Let a constituency have to return 3 Members, and let 5 candidates stand, 3 Liberals, 1 Independent Liberal, and 1 Conservative.

Let 11999 voting-papers be filled up as follows:-

CHAMBERLAIN	4	4	2	1	4	-
GLADSTONE	1	2	1	2	2	-
GOSCHEN	3	3	4	4	1	-
HARTINGTON	2	1	3	3	3	-
NORTHCOTE	-	-	-	-	-	1
Nos of papers	3030	2980	2020	1100	790	2079

Here the necessary 'quota' is 3000, since, if 3 candidates get 3000 each, a fourth can only get 2999.

Also it is clear that, as a matter of justice, Gladstone, Hartington, and Chamberlain ought to be returned, since there are 6010 electors who put Gladstone and Hartington as their first two favorites, and, over and above these, 3120 who put Gladstone and Chamberlain as their first two.

The First Count would give:-

Gladstone	...	...	5050
Hartington	...	...	2980
Northcote	...	...	2079
Chamberlain	...	...	1100
Goschen	...	...	790

Thus Gladstone is returned, with 2050 votes to spare, which must be divided between Hartington and Chamberlain in the proportion of 3030 to 2020, i.e. of 3 to 2; i.e. Hartington must have 1230 of them, and Chamberlain 820.

The Second Count would give:-

Gladstone	...	...	3000
Hartington	...	...	4210
Northcote	...	...	2079
Chamberlain	...	...	1920
Goschen	...	...	790

Thus Hartington is returned, with 1210 votes to spare, the whole of which must go to Goschen.

The Third Count would give:-

Gladstone	...	...	3000
Hartington	...	...	3000
Northcote	...	...	2079
Goschen	...	...	2000
Chamberlain	...	...	1920

What is to be done now? There is one seat yet to be filled, and no one has the necessary quota. Merely counting votes as they now stand would bring in Northcote, which we know would be unfair. The method of the Society in such a case would be (they tell me) to transfer to Goschen so many of Chamberlain's votes as would give him the necessary quota. But this also would bring in the wrong man.

Thus the Society's method not only excludes Chamberlain, who most undoubtedly ought to be returned; but it actually uses, for the purpose of returning Goschen, the votes of 1000 electors who prefer Chamberlain!

May I, in conclusion, point out that the method advocated in my pamphlet (where each elector names one candidate only, and the candidates themselves can, after the numbers are announced, club their votes, so as to bring in others besides those already announced as returned) would be at once perfectly simple and perfectly equitable in its result?

In the above instance the returning-officer would announce as follows:-

"Gladstone	...	5050
Hartington	...	2980
Northcote	...	2079
Chamberlain	...	1100
Goschen	...	790
	4	11999
		2999 and 3/4th[3].

Quota needed to return 1 Member is 3000.
2 .... 6000.
3 .... 9000.

I hereby declare Mr. Gladstone to be duly returned.

Two seats remain to be filled".

Gladstone, Hartington, and Chamberlain would then club their votes, making 9130 votes, which would suffice to return all three.

C. L. DODGSON.

CH. CH., OXFORD
Feb., 1885.

4.1.3. The Principles of Parliamentary Representation. Postscript to Supplement.

OBJECTION has been taken to my statement on page 170 ("it is clear that, as a matter of justice, Gladstone, Hartington, and Chamberlain ought to be returned") on the ground that, of the 9,920 Liberal electors, there are 6,800 who prefer Goschen to Chamberlain, while there are only 3,120 who prefer Chamberlain to Goschen. And it has been pressed upon me that, after all, Goschen is the right man to return, so that the Society's method does *not* break down in this instance.

Now, first, we might almost on *a priori* considerations reject such a test as manifestly unfair. For does it not involve the using an elector's voting-power *more than once?* We first let an elector exhaust his full voting-power in helping to return (say) Gladstone; and, after that, we allow his opinion to have weight in deciding between two other candidates. Is not this to abandon the principle, adopted by the Society, that each elector shall have *one* vote only?

But secondly, this test may be easily proved to be valueless, by a simple *reductio ad absurdum.*

---

3 [Originally 1/4th, corrected in *Postscript to Supplement.*]

Let the 11999 voting-papers be filled up as follows:-

CHAMBERLAIN	4	4	2	1	3	-
GLADSTONE	1	2	1	2	2	-
GOSCHEN	3	3	4	4	1	-
HARTINGTON	2	1	3	3	4	-
NORTHCOTE	-	-	-	-	-	1
Nos. of papers	1826	1712	1826	1712	1910	3013

The First Count would give:-

Gladstone	...	...	3652
Northcote	...	...	3013
Goschen	...	...	1910
Chamberlain	...	...	1712
Hartington	...	...	1712

Thus Gladstone and Northcote are returned, and Gladstone's 652 spare vote must be equally divided between Chamberlain and Hartington.

The Second Count would give:-

Northcote	...	...	3013
Gladstone	...	...	3000
Chamberlain	...	...	2038
Hartington	...	...	2038
Goschen	...	...	1910

What is to be done now? There is one seat yet to be filled, and no one has the necessary quota.

Let us try the new test. And first, of course, we apply it to Chamberlain and Hartington. We find that

5448 prefer Chamberlain to Hartington;
3538 prefer Hartington to Chamberlain;

so that Chamberlain is clearly the victor.

Let us next apply it to Chamberlain and Goschen. We find that

3538 prefer Chamberlain to Goschen;
5448 prefer Goschen to Chamberlain;

so that Goschen is clearly the victor, and we might perhaps rest satisfied that he is the right man to be returned.

Let us however, in order to make assurance doubly sure, apply the test to Goschen and Hartington. We find that

1910 prefer Goschen to Hartington;
7076 prefer Hartington to Goschen!

This lands us in a hopeless circle: and the logical conclusion I believe to be that the proposed test is absolutely valueless.

The statement, thus objected to, may be most fully expressed as follows:-

The 6010 electors, who put Gladstone and Hartington as their first two favorites, have a clear *moral* right to return these two, as they muster more than two full quotas, and it is a mere accident (which they would have avoided had they known how the voting was going on) that they did not divide themselves so as to *secure* this result.

Now suppose them to have exercised this right, and that Gladstone and Hartington are returned. Then the 3120, who put Gladstone and Chamberlain as their first two favorites, would undoubtedly (Gladstone being now safe) all vote for Chamberlain. *He* therefore has also a *moral* right to be returned.

C. L. DODGSON.

CH. CH., OXFORD
Feb., 1885.

4.2. James Garth Marshall: Minorities and Majorities; Their Relative Rights

MINORITIES AND MAJORITIES; THEIR RELATIVE RIGHTS - A LETTER TO THE LORD JOHN RUSSELL, M.P. ON PARLIAMENTARY REFORM.

James Garth Marshall

(London, James Ridgway, 1853)

[Written when electoral reform was at the forefront of the domestic political agenda, the article argues for a change in the system of parliamentary elections, pointing out the misrepresentation arising from the plurality system (or block vote) used at the time. As with most of the literature on electoral reform written in Britain during the nineteenth century, Marshall looks at it in terms of majority and minority representation, echoed in Carroll's use of the two party system in The Principles of Proportional Representation, and argues that 'the mechanism of the elective process should be such as to allow all parties and opinions to be fairly represented in the House of Commons'. Marshall first describes 'the Single Vote' or SNTV, but claims that a far better system, because it better reflects individual preference schedules, would be the Cumulative Vote.

The mathematical analysis, which Black claimed Carroll used as the basis for his exploration of proportional representation, is relegated to an appendix in the pamphlet, and unlike Carroll's work, is not set out in a logical framework, but merely presented as a number of examples of the threshold levels of support required to gain minority representation.]

APPENDIX C

Analysis of the Results of the Cumulative and of the Single Vote in constituencies of 2, 3, 4, or more members, and where the minority is 1/3, 2/5, 1/4th of the whole number of Electors.

The cases examined are:-

I.	Two Members. Three Candidates	1. Minority 1/3rd.
		2. Minority 2/5th.
		3. Minority 1/4th.
II.	Two Members. Four Candidates	4. Minority 1/3rd.
		5 & 6. Minority 2/5th. & 1/4th.
III.	Three Members. Four Candidates	7. Minority 1/3rd.
	Three Members. Five or Six Candidates	8 & 9. Minority 2/5th.
IV.	Four Members. Five, Six, Seven, Eight Candidates.	
V.	The Single Vote in all the above cases.	

The Constituency in every case is supposed to consist of 300 Voters, and the object of the Analysis is to ascertain whether the minority obtain, in all cases, a proportionate share, and no more than a proportionate share of the representation.

I.

2 Members; 3 Candidates; minority 1/3rd = 100.

The minority have 200 votes, the majority 400. It is an even chance whether the minority carry 1 member, or whether the majority carry 2. If the minority *exceed* 1/3, they have half the Representation, if they are less than 1/3 they have no Representation.

(2.) 2 Members; 3 Candidates; minority 2/5th = 120

In this case the minority are secure of one member. They cannot carry 2. The majority are secure of 1 member, but cannot carry 2, unless the minority have 2 candidates, which is not the supposition.

But suppose a split in the majority, an the smaller section, by aid of surplus votes of the minority, secure their member, causing the larger section of the majority to be unrepresented? The majority 180 = 360 votes. The minority 120 = 240 votes. Then minority 120 + 81, smaller section of majority, 201 = 402 votes. And larger section of majority, 99 = 198 votes.

The result is that if so many of the majority joined the minority as to make their number exceed 2/3 of the whole votes, they would carry 2 members, and exclude the larger section of the majority reduced to less than 1/3. This seems to be fair. But how would it be on the present mode of voting, plumpers being single votes?

Minority 120 + 31, smaller section of majority, 151 = 151 votes each for 2 members. And larger section of majority = 149 plumpers.

The result is, that if so many of the majority joined the minority as made them barely exceed 1/2 the voters, they would carry 2 members, and the larger section of the majority, barely less than 1/2 the voters, would be entirely unrepresented.

So that the Cumulative Vote is in this case much more fair towards a divided majority, and less likely to induce factious splitting of votes than the present mode of voting.

(3.) 2 Members; 3 Candidates; minority, 1/4 = 75.

In this case, as the minority have no chance when the majority are united, we must suppose a split in the majority, and see how far the minority can pick out and secure 1 member most to their liking.

Minority, 75 + 26, smaller section of majority, 101 = 202 votes.

Larger section of majority, 199 = 398 votes = 199 votes for each of 2 members.

So that here again we turn on the same point. If so many of the majority join the minority as to make them *more* than 1/3, they secure 1 member, which is fair. On the present mode of voting, the minority of 1/4th would be entirely swamped unless so many of the majority split with them as to make them more than 1/2 the voters, and then they would carry 2 members, and the larger section of the majority would be entirely swamped.

In this case, too, the "Cumulative Vote" seems more equitable than the present mode; tends less to injustice or factious splittings.

II.

2 Members; 4 Candidates; minority, 1/3 = 100.

(4.) In this case we must suppose the majority or minority to have each 2 candidates; and if each party is united the majority must carry both members.

(5) and (6). And the result would be similar when the minority is 2/5 or 1/4. When each party brings out 2 candidates it is reasonable to suppose they are both united within themselves; and then the minority have no chance, aiming at too much.

III.

3 Members; 4 Candidates; minority, 1/3 = 100.

Minority have 300 votes; majority, 600.

Here, if both parties are united within themselves the minority are secure of 1 member, but cannot carry 2; but, suppose a split in the majority, can the smaller section, by aid of the surplus votes of the minority, carry their 1 or 2 candidates, excluding 1 or 2 supported by the larger section?

In this case the smallest number of votes to secure a candidate is 226.

	Candidates			
	Minority		Majority	
	A	**B**	**C**	**D**
Minority, 300 votes	226	74		
Majority, 600 votes		148	226	226
	226	222	226	226

This shows, that so long as the larger section of the majority *exceed* 2/3 of their *own* party (or 1/2 the whole number of votes - in this case 450 votes) they can secure 2 members. But if they are *less* than 2/3 of their own party and less than 1/2 the total votes, they can only secure 1 member.

It may, indeed, appear at first sight that the majority might divide their votes equally amongst 3 candidates, 200 each; and in that case the minority, by dividing their 74 spare votes between 2 of the other candidates, might secure both. But such a case would be imaginary merely, and could not happen in practice. The majority, in bringing out 3 candidates, must know they were liable to this occurrence; there must be a split to begin with, or they would be content with 2 candidates. The split being known and acted upon, the result would be as first stated above.

(8) and (9). 3 Members; and 5 or 6 candidates; minority, 2/5 = 120.

Minority, 360 votes; majority, 540 votes.

This case will be easily understood by reference to the last. With 5 candidates, 181 is the smallest number of votes to secure a seat.

	Candidates				
	Minority		Majority		
	A	B	C	D	E
Minority, 360 votes	180	180			
Majority, 540 votes	1	1	181	178	179
	181	**181**	**181**	**178**	**179**

With six candidates, 151 is the smallest number.

	Candidates					
	Minority			Majority		
	A	B	C	D	E	F
Minority, 360 votes	151	151	58			
Majority, 540 votes			93	149	149	149
	151	**151**	**151**	**149**	**149**	**149**

In both cases the minority is secure of 1 member, and the majority is secure of 2 members if each party is tolerably united within itself.

The above tables show how a minority may possibly carry 2 out of 5, or 3 out [of] 6 candidates, if aided by a section of the majority.

But it is only necessary to consider the numbers to see that the majority must act with so little unity within themselves, and so much in alliance with the minority in order to give the minority their 2 or 3 members that it virtually amounts to a conversion of the minority into a majority. In practice it could rarely happen, that in the case of 5 candidates the two first of the majority should not poll more than the second candidate of the minority; or that in the case of six candidates they should divide their votes so equally amongst 4 candidates. If they do so divide them, it shows that in such case no strong general party feeling exists, but that the candidates are supported for local or personal reasons. Now that is usually the defect of small constituencies; but the opposite character is usually the defect of large, and particularly large town constituencies. There too little weight is apt to be given to personal character and to local considerations, and general party spirit runs too high. And it is an advantage of the Cumulative Vote that it would tend to correct this defect; but the present mode of voting tends to exaggerate it. Where three or four members are to be chosen for an important constituency a simple majority carries the whole. The motives of intense party feeling are maximised.

IV.
4 Members; and 5, 6, 7, 8 Candidates

It seems to me scarcely necessary to work out these calculations in detail further. I can see nothing in the principle of the Cumulative Vote as already analyzed to cause it to act otherwise than fairly both to majority and minority; if we can only divest ourselves of the traditional prepossession that the majority have, somehow or other, a right to more than their share, and are ill used if they do not get it.

V.
The single vote plan. All constituencies to have three members, and each voter only one vote.

Candidates 4, 5, or 6; minorities 1/4, 1/3, 2/5.

In this mode of voting, out of a constituency of 300 votes, the minorities, 1/3, 1/4, 2/5 - that is; 75, 100, 120 respectively - would have, it seems to me, exactly the same absolute weight compared to the majority, the same means of taking advantage of a split in the majority, as in the cases we have analyzed where each voter has 3 votes. The difference in practice would be, that whereas, in order to carry out party tactics, there would be, on the plan to the Cumulative Vote, a splitting of votes, so on the single vote plan there would be a splitting of voters, so to speak. That is, the voters in each party must make an arrangement amongst themselves, that so many will vote for candidate A, and so many for candidate B, &c.. Otherwise there would be no combined party action; each candidate would stand separately, and there might be an indefinite number of candidates.

It is reasonable to suppose that party action would not cease; and that each party would endeavour so to split and divide their voters as to carry as many members as they can. If so I can see no advantage in the single over the Cumulative Vote: we reach the same final result by less simple means, for I think it will be conceded this splitting or distribution of voters would be less easy and simple than the splitting of votes.

4.3. Walter Baily: Proportional Representation in Large Constituencies

PROPORTIONAL REPRESENTATION IN LARGE CONSTITUENCIES

By Walter Baily

(London, Ridgway, 1872)

Among the questions of the present day there is none which has attracted a more rapidly increasing or widely spreading interest that the question of the best mode of electing a body of Representatives. Publications on this question issue from the presses not only of England, but also of America, Germany, France, Italy, Sweden, Denmark, and Switzerland. The old system which allowed only the majority of a constituency to elect representatives, and shut out minorities from all share in the governing body, meets with general condemnation, and a new scheme is sought for which shall secure as far as possible to every vote in a constituency an equal share in a representative, and so produce a representative body in which the various parties shall have as nearly as possible the same proportionate strength that they have in constituency itself. This is what is called 'Proportional Representation'.

A great practical step in the direction of proportional Representation was made when 'Cumulative Voting' was introduced by the Elementary Education Act, 1870, into the election of School Boards for the purpose of giving representation to conflicting parties, and the results of the elections held under this Act produced so much satisfaction that an attempt made the following session to bring about a return to the usual manner of voting met with complete failure. Nevertheless 'Cumulative Voting' is a very imperfect system, not capable of giving representation in due proportion, except by a happy accident, and sometimes giving it very unfairly, as the example of the Marylebone School Board election will show. Out of 165,115 votes given to elect seven members, 47,858 votes are represented by *one* member only, 63,791 votes are represented by the remaining *six* members, and 53,466 votes are not represented at all.

The causes of failure are easy to see. Some votes are wasted by being given to candidates who have enough without them, and others are wasted by being given to candidates who are not elected. A sound system of election must provide, as much as may be, against both kinds of waste. This can be done by the contingent or shifting vote advocated by Mr. Hare. His scheme may be shortly described as follows:-

Each elector has one vote, and gives to the polling clerk a list of candidates in the order of his choice; the number of these lists is ascertained, and then the vote of each elector is assigned to the candidate whose name heads his list. As soon as a candidate has obtained a sufficient number of votes, no more are given to him; but when his name occurs on a list, it is passed over in favour of the next name on the list. Provision is thus made against the waste of votes on popular candidates. When all the votes have been distributed, there will generally be some candidates with sufficient votes, and others with smaller numbers of votes. If we declare elected those with sufficient votes, and as many others as may be wanted to complete the number of members, taking of course those with the greater number of votes, we have the plan actually in use in Denmark; but it is obviously defective, as it does not provide against the waste of votes on unsuccessful candidates.

Here the mode of electing a preacher at Lincoln's Inn may be usefully noticed. A ballot is taken, and results naturally in several candidates getting votes, but perhaps none getting an absolute majority. In that case some of the lowest candidates are withdrawn and another ballot taken, more are withdrawn, and if needful a third ballot taken, more are withdrawn, and if needful a third ballot taken, until at last a candidate is elected by an absolute majority.

To apply this device to the election by lists, we should have, when all the votes are distributed, to withdraw the lowest candidate, and distribute his votes among the remaining candidates according to the lists, then to withdraw another candidate and distribute his votes, and so on until the election ended either by the right number of candidates having got sufficient votes, or by all except the right number having been withdrawn. Provision has now been made as far as possible against the waste of votes upon unsuccessful candidates.

We have still to consider what is the sufficient number of votes to be retained for each candidate. The rule in use in Denmark, and adopted by Mr. Hare, for finding this number, which is called the Quota, is to divide the number of votes by the number of members to be elected. This is simple, but still it is wrong. for example, if we apply Mr. Hare's plan to an election of two members in which 100 votes are given, 70 for A first, and then B, and 30 for C, we should obtain the Quota by dividing 100 by 2, and then retaining this Quota of 50 votes for a, we should hand over 20 votes to B, and the votes would then stand A 50, C 30, B 20, and therefore we should have A and C elected; and yet it is clear that, as 70 is more than twice 30, A and B should have been the candidates elected.

The number of votes to be retained for a candidate must be enough to make his election certain, whatever combination may be made of the other votes given in the election; the smallest number which will suffice for this is the true Quota; all votes retained beyond this number are wasted. There is no difficulty in finding this number. Suppose that two members have to be elected, we must retain for a candidate votes enough to insure his being one of the first two, and this we shall do if we retain for him just over a third of the whole number of votes given. It is impossible for three persons, each to have more than one-third of the votes, so that any candidate who has more than one-third by ever so little, is certain to be one of the first two, in whatever way the rest of the votes may be distributed. In the same way we see that if five members have to be elected, a candidate who has more than one-sixth of the votes, will certainly be one of the first five, and therefore elected; and so for any other number of members. The rule, then, for finding the true Quota is this: Divide the number of votes by the number just above that of the members to be elected, and take as the Quota the number just above the Quotient.

In the example given above, the true Quota just exceeds one-third of 100. It is therefore 34. The 70 votes given to A, B, will then be divided into 34 for A, 34 for B, and 2 over. C has only 30 votes; and the result is that A and B are elected, as it is clear they should be.

It will be observed that some votes are wasted. This must needs be whatever mode of election is adopted. If a constituency has only one member, a candidate who gets a bare majority will be elected, and it will be of no moment whether the remaining votes are for him or against him. All except the bare majority can have no effect upon the election, and may be considered as wasted. But as the number of members is increased, the unavoidable waste is diminished. With five members, the effective votes for each will just exceed one-sixth, and therefore the waste votes will just fall short of the remaining sixth. In fact, the unavoidable waste will always just fall short of the true Quota.

In the Marylebone School Board election already referred to, as seven members had to be elected, the true Quota just exceeded one-eighth of the 165,115 votes, so that it was 20,640. The waste votes should have been just under this number, but they really amounted to 53,466 votes given to unsuccessful candidates, and 27,218 votes given to the most popular candidate in excess of the Quota, in all 80,684 votes, or nearly half the number of votes given in the election.

Mr. Hare's scheme with the alteration discussed above, would stand thus:-

Each voter gives in a list of candidates.

The lists are counted, and the true Quota calculated.

The votes are distributed according to the lists, but no more than the Quota are retained for any candidate.

The lowest candidates are withdrawn one by one, (if two are equal, one is taken by lot), and their votes distributed among the remaining candidates.

The result of an election conducted on this plan would be that the vote of every voter would be that the vote of every voter would be used for the earliest name on his list, for which it could be made effective, and (if the voters did not make their lists too short) every member would be elected by a full Quota of votes, and the number of ineffective or waste votes would be the smallest possible.

[Baily continues, suggesting that the Hare, or STV, system outlined above is too complicated to use in large constituencies with a large, and ill-educated, electorate. As an alternative he proposes a simplified system in which the candidate indicates how any surplus votes he gets will be transferred, similar to a party list system of conducting elections. Instead of having to indicate relative preferences between all

the candidates nominated, the electorate is presented with a number of predetermined lists, surplus votes for the leading candidate are transferred to the next candidate on the list, and so on. The list system would work in the same basic way as the Hare system, aiming to keep the number of wasted votes as small as possible, just under one complete Quota. This system is essentially the same as Carroll's, with candidates determining where their surplus votes would be distributed.]

Index

A

B

C

D

E

F

G

H

I

K

L

M

N

O

P

Q

R

S

T

U

V

W

Z

MIX
Papier aus verantwortungsvollen Quellen
Paper from responsible sources
FSC® C105338

If you have any concerns about our products,
you can contact us on
ProductSafety@springernature.com

In case Publisher is established outside the EU,
the EU authorized representative is:
Springer Nature Customer Service Center GmbH
Europaplatz 3, 69115 Heidelberg, Germany

Printed by Libri Plureos GmbH
in Hamburg, Germany